74th Annual Meeting of the Northeastern Weed Science Society 2020

Proceedings – Volume 74

Philadelphia, Pennsylvania, USA
6 – 10 January 2020

Editor:

Michael L. Flessner

ISBN: 978-1-7138-0873-2

Printed from e-media with permission by:

Curran Associates, Inc.
57 Morehouse Lane
Red Hook, NY 12571

CURRAN ASSOCIATES INC. proceedings.com

Some format issues inherent in the e-media version may also appear in this print version.

Northeastern Weed Science Society (NEWSS)
P.O. Box 25
Woodstown, NJ 08098
USA

Phone: 315-209-7580

northeasternweedscience@hotmail.com

Additional copies of this publication are available from:

Curran Associates, Inc.
57 Morehouse Lane
Red Hook, NY 12571 USA
Phone: 845-758-0400
Fax: 845-758-2633
Email: curran@proceedings.com
Web: www.proceedings.com

PROCEEDINGS - VOLUME 74

74th Annual Meeting of the

Northeastern Weed Science Society

January 6-10, 2020 – Hilton Philadelphia at Penn's Landing
Philadelphia, PA

Site of 2021 Joint Meeting*

January 4-7, 2021

Wyndham Gettysburg
Gettysburg, PA

Participating Societies:
Northeastern Weed Science Society
Northeastern Branch – American Society of Agronomy/Crop Science Society of America/Soil Science Society of America
American Society of Horticultural Science – Northern Region

§ Cover: The winning image in the 2020 NEWSS Photo Contest, submitted by Hilary A. Sandler, University of Massachusetts Cranberry Station, Dodder on Cranberry.

Proceedings
of the
Seventy Fourth Annual Meeting
of the
Northeastern Weed Science Society

Michael L Flessner, Editor

2020 SUSTAINING MEMBERS

Platinum

Gold

Silver

Bronze

Nufarm	AMVAC
LABServices	For-Shore Weed Control

Table of Contents

NEWSS EXECUTIVE COMMITTEE – OFFICERS

President	**Art Gover** The Pennsylvania State University aeg2@psu.edu
President-elect	**Caren Schmidt** BASF caren.schmidt@basf.com
Vice President	**Jacob Barney** Virginia Tech jnbarney@vt.edu
Treasurer	**Kurt Vollmer** University of Maryland kvollmer@umd.edu
Treasurer Elect	**Larissa Smith** Syngenta Larissa.smith@syngenta.com
Past President	**Dan Kunkel** Rutgers University daniel.kunkel@rutgers.edu

NEWSS EXECUTIVE COMMITTEE – MEMBERS

Membership Chair	**Thierry Besançon** Rutgers University thierry.besancon@rutgers.edu
Editor	**Lane Heimer** Maryland Department of Agriculture lane.heimer@maryland.gov
Public Relations	**Katelyn Venner** BASF katelyn.venner@basf.com
Research and Education Coordinator	**Matt Cutulle** Clemson University mcutull@clemson.edu
Sustaining Membership	**Wesley Everman** North Carolina State University wes_everman@ncsu.edu
CAST Representative	**Hilary Sandler** University of Massachusetts Cranberry Station hsandler@umass.edu
Graduate Student Representative	**Uri Menalled** Cornell University udm3@cornell.edu
Graduate Student Representative Elect	**Maria Gannett** Cornell University mag449@cornell.edu
WSSA Representative	**Rakesh Chandran** West Virginia University rschandran@mail.wvu.edu
Science Policy Director	**Lee Van Wychen** Weed Science Society of America lee.vanwychen@wssa.net

2020 NEWSS MEETING PROGRAM COMMITTEES

Program Committee	Chair:	Caren Schmidt, BASF
Agronomic Crops	Chair:	John Wallace, The Pennsylvania State University
Turfgrass, Ornamentals, and Non-Crop	Chair:	Jatinder Aulakh, Connecticut Ag Experiment Station
Fruit, Vegetable, and Herb Crops	Chair:	Thierry Besançon, Rutgers University
Research Posters	Chair:	Kate Venner, BASF
Student Oral Papers	Chair:	Carroll Moseley, Syngenta
Student Posters	Chair:	Anthony Witcher, Tennessee State University

2020 NEWSS COMMITTEES

Student Paper Contest	Chair:	Daniel Kunkel, Rutgers University
Student Poster Contest	Chair:	Anthony Witcher, Tennessee State University
Photo Judging Contest	Chair:	Hilary Sandler, University of Massachusetts
Site Selection Committee	Chair:	Caren Schmidt, BASF
Awards Committee	Chair:	Daniel Kunkel, Rutgers University
Nomination Committee	Chair:	Carolyn Lowry, The Pennsylvania State University
Resolutions Committee	Chair:	Jerry Baron, Rutgers University
Collegiate Weed Contest	Chair:	Caren Schmidt, BASF
Archives Committee	Chair:	Daniel Kunkel, Rutgers University
Weed Science Field Days	Chair:	Quintin Johnson, University of Delaware
Past Presidents Committee	Chair:	Daniel Kunkel, Rutgers University
Herbicide Resistance	Chair:	Mark VanGessel, University of Delaware
Endowment Committee	Chair:	Jacob Barney, Virginia Tech

2020 NEWSS AWARDS

NEWSS Fellow Award – Tommy Hines

NEWSS Fellow Award – Grant Jordan

NEWSS Fellow Award – Dwight Lingenfelter

Award of Merit – Dave Mayonado

Award of Merit – Dave Jim Steffel

Outstanding Researcher Award – *none awarded*

Outstanding Educator Award – Jacob Barney

Robert D. Sweet Outstanding Ph.D. Student Award – John Brewer

Robert D. Sweet Outstanding M.S. Student Award – Lucas Rector

M. Garry Schnappinger Service Recognition Award – Rakesh Chandran

2020 NEWSS MEETING CONTEST WINNERS

STUDENT PAPER

1st place paper – Katie Diehl, Rutgers University, "Annual Bluegrass Weevil (*Listronotus maculicollis*), Overseeding, and Paclobutrazol for Annual Bluegrass (*Poa annua*) Control in Fairways."

2nd place paper – Maggie Wasacz, Rutgers University, "Evaluating Crop Response of Various Vegetable Species to Dicamba Drift."

Honorable mention – Vasiliy Lakoba, Virginia Tech, "Ecotypic Stress Responses in an Invasive Grass are Driven by Home Climate and Habitat."

STUDENT POSTER

1st place: Eric Jones, North Carolina State University, "Investigating Potential Antagonism Between Sequential Applications of Dicamba and Glufosinate."

2nd place: Eugene Law, Cornell University, "Scientific Literacy in the Weed Science Classroom."

3rd place: Theresa Reinhardt Piskackova, North Carolina State University, "Can Repeat Photography Be Used to Generate Weed Emergence Models?"

PHOTO CONTEST

1st place photo – Hilary A. Sandler, University of Massachusetts Cranberry Station, Dodder on cranberry

2nd place photo – Thierry Besançon, Rutgers University, Meadow Beauty

3rd place photo - Thierry Besançon, Rutgers University, Tickseed sunflower

2019 NORTH AMERICAN WEED SCIENCE CONTEST

BASF Midwest Research Farm, Midwest Ag Research Center of Valent, and Klein Farms near Seymour, IL

Hosted by: BASF, Valent, Corteva and Bayer

(Photos courtesy of Katelyn Venner)

The 2019 North American Weed Science Contest, affectionately known as the "Weed Olympics", was hosted at the BASF Midwest Research Farm, Midwest Ag Research Center of Valent, and Klein Farms near Seymour, IL on July 25, 2019. Since the "Weed Olympics" is a national event, students from The Northeastern Weed Science Society, Southern Weed Science Society, North-Central Weed Science Society, and Western Weed Science Society participated in the contest. A total of 63 teams representing 26 graduate programs and 8 undergraduate programs competed in this rigorous contest. The day of the contest was sunny, beautiful, and ripe for competition. The contest concluded with the awards ceremony at the University of Illinois Memorial Stadium, where students were treated to a fantastic view of the stadium facility. The contest was quite competitive, Virginia Tech Graduate Team 1 won the overall graduate student division and took first place in the NEWSS section of the contest. Virginia Tech Graduate Team 2 took third place in the NEWSS section of the contest. University of Guelph Graduate Team took second place in the NEWSS section, and their Undergraduate Teams took first and third place in the NEWSS undergraduate division. The hosts and sponsors of this years "Weed Olympics" put on a fantastic contest, and the students and coaches probably can't wait for the next one!

Workin' Hard on the Written Exam

What's the Farmer's Problem?

The Medal Ceremony

Overall North American Weed Science Contest Awards

(as they pertain to NEWSS)

Overall Winner – Individual (Graduate)

Eric Scruggs

Virginia Tech

Overall Winner – Individual (Undergraduate)

Matt Spoth

Cornell University

Unknown Herbicide (Graduate)

Eric Scruggs

Virginia Tech

Weed Identification (Graduate)

John Brewer

Virginia Tech

Problem Solving (Undergraduate)

Kurtis Pilkington

University of Guelph

Overall Contest Winners

Virginia Tech Graduate Team 1 took home the gold at the North American Weed Science Contest.

Left to right: Wykle Green, Eric Scruggs, John Brewer, Shawn Beam

NEWSS Undergraduate Team Awards

First Place Undergraduate

University of Guelph Undergraduate Team

Matt Fletcher, Aleksander Kus, Hannah Symington, Kurtis Pilkington

Second Place Undergraduate

Cornell University Undergraduate Team

Aleah Butler-Jones, Emma Kubinski, Matt Spoth

Third Place Undergraduate

University of Guelph Undergraduate Team

Emily Duenk, Bryce Ribey, Alan Abdulkader, Elaine Jeffs

Undergraduate Individual Awards

First Place Undergraduate

Matt Spoth

Cornell University

Second Place Undergraduate

Kurtis Pilkington

University of Guelph

Third Place Undergraduate

Bernard Wierenga

University of Guelph

Graduate Student Team Awards

First Place Graduate Team

Virginia Tech Graduate Team

John Brewer, Shawn Beam, Wykle Greene, Eric Scruggs

Second Place Graduate Team

University of Guelph Graduate Team

Alex Sanders, Brett Hilker, Allison Bailey, Bernard Wierenga

Third Place Graduate Team

Virginia Tech Graduate Team

Gourav Sharma, Jordan Craft, Spencer Michael, Caleb Henderson

Graduate Individual Awards

First Place Graduate Student

Eric Scruggs

Virginia Tech

Second Place Graduate Student

Shawn Beam

Virginia Tech

Third Place Graduate Student

John Brewer

Virginia Tech

ROW SPACING AND PLANTING DATE EFFECT ON THE CRITICAL PERIOD OF GRASS WEED CONTROL IN GRAIN SORGHUM. D.J. Contreras* and W.J. Everman, North Carolina State University, Raleigh, NC

ABSTRACT

Grain sorghum (*Sorghum bicolor*), which is mainly used for animal feed, is the 4th most planted crop in the US. Its production is focused in the Midwest, where precipitation is limiting for high yielding corn and soybean production. North Carolina growers can benefit from grain sorghum production, especially in dry and hot areas, due to a constant demand for animal feed. Weed management is a concern for grain sorghum growers due the limited number of herbicide options available. Grass weed species have been able to thrive under conditions were most herbicide control programs are focused on broadleaf weeds because of lack of selective annual graminicides. Cultural control practices such specialized crop varieties with herbicide tolerance, row width variations, planting date and Critical Period of Weed Control (CPWC) are some of the tools available for weed management. The objective of this study was to determine the CPWC for grass species in grain sorghum using a crop variety with Inzen™ Sorghum Trait technology, two different row width arrangements (15 and 36 in.) and two planting dates (Mid-May and Mid-June). Plots were kept weedy or weed free through 2, 3, 5, and 7 weeks after crop emergence (WAE). Two controls consisting of weedy and weed free all season plots were used for a treatment comparison based on a regression analysis where time of weed removal was related to crop yield. An acceptable yield loss of 5% was given to limit the CPWC. Grain sorghum planted in Mid-May had a CPWC that lasted from 2.5 to 7.6 WAE for 15" row spacing and from 2.2 to 7.8 WAE for 36" row spacing ($p \leq 0.05$). Grain sorghum planted in Mid-June had a CPWC from 3.4 to 6.7 WAE for 15" row spacing and 2.9 to 6.3 WAE for 36" row spacing ($p \leq 0.05$).

POST-HARVEST WEED COMPETITION IN PERENNIAL GRAIN CROPS: A NEW CRITICAL PERIOD OF WEED CONTROL? M.P. Spoth*, E.P. Law, M.R. Ryan, S. Wayman, C.J. Pelzer, and A. DiTommaso, Cornell University, Ithaca, NY

ABSTRACT

Perennial grain crops such as 'Kernza®' intermediate wheatgrass (*Thinopyrum intermedium*; IWG) and 'ACE-1' perennial cereal rye (*Secale cereale* x *S. strictum*; PR) have the potential to contribute to the sustainability of diversified cropping systems due to their ability to provide ecosystem services beyond food and forage production but suffer economically from relatively low grain yields. One area of agronomic management of perennial grains that has not been thoroughly examined is the impact of weed competition on perennial grain growth and productivity. Weed management has been a consistent issue during establishment of IWG in both field experiments and on-farm trials in New York. In contrast, PR has had few issues with weeds during establishment but appears to be severely impacted by weed competition after harvest. Here we report on a preliminary study designed to provide insight into the impact of the timing of weed control interventions in the period between the first and second harvests of these two crops implemented at Musgrave Research Farm in Aurora, NY. Subplots of IWG, PR, and both species intercropped with field pea (*Pisum sativum*) were subjected to four hand weeding treatments applied at different times: every two weeks from harvest to winter dormancy ('Fall'), from spring greenup to anthesis ('Spring'), both Fall and Spring, and an unweeded Control. Crop and weed biomass were sampled from 0.5 m^2 quadrats at grain maturity in July and August 2019. Yields of IWG were highest when both intercropped with field pea and weeded in Fall and Spring compared to the Control weeding treatments, despite that treatment combination producing one of the lowest amounts of total crop biomass. In contrast, PR yields were highest when grown in monoculture and weeded in the Spring or Fall and Spring; in general PR/pea intercrops performed poorly. These results will help target future research into crop/weed competition in perennial grains and have the potential to influence management recommendations for timing of weed control interventions.

MICROBIAL CONTRIBUTIONS TO WEED SUPPRESSION IN CONVENTIONAL AND ORGANIC FARM SOILS. A. DiTommaso, J. Kao-Kniffin, and L. CHENG*, CORNELL UNIVERSITY, Ithaca, NY

ABSTRACT

[no abstract entered]

CEREAL RYE SEEDING RATE EFFECTS ON WEED SUPPRESSION IN PLANTING GREEN SOYBEANS. T. Ficks*, J. Wallace, and D. Lingenfelter, Penn State, State College, PA

ABSTRACT

In agronomic no-till cropping systems, there is an increased interest in planting green to improve soil health. Planting green is a management practice where cover crop termination is delayed until at or after cash crop planting. Growers use different cover crop species, seeding rate, termination-timing, and roll-plant tactics when adopting this practice. We are interested in exploring the effects of planting green tactics on weed population dynamics and their impact on herbicide resistance management goals. We conducted a field experiment at Rock Springs, PA (RELARC) in 2018-19 using a RCBD with four replicates. Treatments included four cereal rye seeding rate: 0, 51, 101, 135 kg/ha. Dicamba-tolerant soybeans were planted 1 d prior to a cover crop burndown herbicide application (glyphosate + dicamba) using a no-till planter equipped with integrated ZRX rollers. Aboveground cereal rye biomass was harvested one week prior to termination. Horseweed (*Erigeron canadensis*) density and height were recorded just prior to planting and a post-emergent (6 WAP) herbicide application. Large crabgrass (*Digitaria sanguinalis*), redroot pigweed (*Amaranthus retroflexus*), and common lambsquarters (*Chenopodium album*) density and height were recorded at post-emergent herbicide application timing within established artificial weed seedbanks. Weed biomass was collected in mid-August. Analysis of variance was used to assess treatment effects in R. Excluding the fallow control, rye biomass at termination did not differ across seeding rate treatments. Horseweed and summer annual weed densities were significantly reduced in all treatments compared to the fallow control. The distribution of redroot pigweed heights at the time of post-emergent applications included individuals that exceeded 15 cm in each treatment with exception of the 101 kg/ha seeding rate. Total late season weed biomass was significantly lower in seeding rate treatments compared to the fallow control treatment but did not differ in pairwise comparisons. Our first-year results suggest that planting green tactics can contribute to proactive herbicide resistance management goals even at low, and economically feasible, cereal rye seeding rates. Future work will include replication of the experiment and a complementary field experiments to further contrast the impacts of planting green tactics on herbicide resistance management goals.

ADAPTING A CONTROLLED ENVIRONMENT MOISTURE STRESS PROTOCOL FROM PERENNIAL TO ANNUAL CROPPING SYSTEM. M. Granadino*, W.J. Everman, G.M. Henry, J.T. Sanders, E.A. Jones, and M.A. Fajardo, North Carolina State University, RALEIGH, NC

ABSTRACT

Drought has been one of the most important limiting factors for crop production, which deleteriously affects food security worldwide. Water is extremely important for crop production. When it becomes limiting to the plant it is important to understand how the plant uses its available water. Vegetative growth on soybean during drought is diminished. Drought stressed soybeans are often shorter with smaller leaves due to a lack of water, nutrient availability and nutrient uptake. Thus, the objective of this study is to adapt a controlled environment moisture stress protocol used in perennial cropping systems, developed by Mueller-Dombois and Sims (1966), to annual cropping systems to better evaluate annual crop behavior under drought conditions. Four water table depth gradient tanks were constructed in a greenhouse in the Central Crops Research Station in Clayton, NC. Each tank was steeply sloped and had a volume of nearly 4 m^3. Soybeans were planted 12 cm inches apart in 9 rows (water levels), 8 plants per row. Plants were grown to V3 growth stage, at this point underground irrigation was turned on. Underground irrigation was supplied through a PVC pipe placed at the backside of the tank; a standpipe was placed at the front of the tank to release excess water. At V4 overhead irrigation changed to 4 water regimes: No irrigation, once a day, twice a day and once every two days; 210 ml per irrigation time was supplied. Moisture content throughout the tanks were taken through four moisture sensors buried randomly through the tank. Weekly heights and volumetric water content (VWC) were taken for three weeks after V4. On week three plants were randomly removed from each row the following measurements were taken: Number of leaves, final height, shoot and root weight, root length. VWC at 12 and 20 cm gradually decreased from levels 1 through 9. Water levels 1-3 presented the largest shoot heights, while levels 7-9 had the lowest. Both dry and fresh shoot weights were highest on the no irrigation water regime which could indicate an increased concentration of lignin in the plant tissue. Root lengths were found to be longer at lower water availability. Root weight was greater on the lower levels. Shoot height to root length ratio gradually decreased as the water levels increased indicating focus on root elongation in search of water as availability decreased. Water table gradient tank protocol successfully demonstrated a gradual increase in drought stress on soybeans.

CAN REPEAT PHOTOGRAPHY BE USED TO GENERATE WEED EMERGENCE MODELS? T.A. Reinhardt Piskackova*, S. Reberg Horton, R.J. Richardson, K.M. Jennings, R. Austin, and R.G. Leon, North Carolina State University, Raleigh, NC

ABSTRACT

Precision timing for integrated weed management requires predictive weed models, but these often are so tedious that the models are based on small area in few locations. Image analysis might increase the frequency and expanse of data collected for model creation. While image analysis rarely has yielded accurate prediction of seedling quantity, we hypothesize that the pattern of emergence could be predicted using relative cumulative pixels from repeat images collected over a season. In this study, image classification was used to determine number of pixels classified as the emerging weed over time. The cumulative number of pixels was used to create predictive models which were compared to the true counts. The models created using images were able to predict emergence patterns from true counts. These results imply that models created from images could be sufficient to model emergence and allow incorporation of larger areas to account for more location variability.

SCIENTIFIC LITERACY IN THE WEED SCIENCE CLASSROOM. E.P. Law*, K.M. Averill, and A. DiTommaso, Cornell University, Ithaca, NY

ABSTRACT

Scientific literacy is an oft-stated learning objective for students from kindergarten through advanced degrees, defined broadly as the skills required to find, understand, and assess the validity and impact of scientific literature and other media. College students pursuing degrees in Science, Technology, Engineering, and Mathematics (STEM) fields, such as weed science and other agricultural sciences, are often assumed to possess these skills based on their previous coursework, and are expected to be able to locate, read, and interpret sources as part of their curriculum. In the experience of the authors, however, few upper-level STEM courses include explicit instruction on how to do so, and there have been few reports of formal assessments of students' experiences engaging with the scientific literature that can be used to inform best practices for literacy interventions at this level. Here we report on students' self-reported proficiency and confidence in reading and writing primary scientific literature in a weed science course at a prestigious US university, and correlate their responses with an assessment of their competence in these areas as demonstrated by their performance on scientific writing assignments. All respondents ($n = 8$, primarily third-year agricultural science undergraduates) believed scientific journal articles were "Hard to read, but worth the effort" both before and after completing the course. On average students had been asked to read primary scientific literature in 3.6 previous courses, had been asked to discuss the literature in 2.3 previous courses, and had been asked to write in the style of a scientific journal article in 2.3 courses. Due to small sample size no responses changed (at an α of 0.05) from the pre- to post-surveys, but several interesting trends with marginal significance were noted. Students indicated that the influence of journal articles on their understanding of science was slightly below "Moderate" (2.88/5) before the course and slightly above that (3.57/5; $p = 0.076$) after the course. Student confidence in reading, writing, and interpreting scientific papers was also "Moderate" (3.12/5) prior to the course and slightly higher on average (3.58/5; $p = 0.107$) after. Further analysis of student writing assignments. We will be conducting further analysis of students' written assignments from the course and intend to repeat these assessments in future years to develop a more robust dataset that will hopefully provide more insight into the impact of these science literacy interventions.

EFFECTS OF SUB-LETHAL DICAMBA RATES ON DIFFERENT VARIETIES AND MATURITY GROUPS V & VI SOYBEANS. M.A. Fajardo* and W.J. Everman, North Carolina State University, Raleigh, NC

ABSTRACT

Soybean (Glycine max) production locate USA as the leader in terms in production worldwide and the second country with the highest exports of the grain. Animal protein feed and vegetable oil production are the main final processing products of soybeans. Different varieties and maturity groups adapted to the different regions of the country had become in greater yield productions along past years. Dicamba tolerant beans have created the opportunity to farmers to improve weed management and reducing the injury rates of their production fields. Along this the potential of increasing other susceptible crops injury has risen. Defining the different effects of sub lethal Dicamba rates, when exposed at vegetative (V4) or reproductive stages (R2) and with two different planting dates (May or June) in terms of injury and subsequently in yield was the objective of this research. Two different maturity groups (V and VI), each with Dicamba tolerant and non-tolerant soybean varieties were used. The location of the research was the Upper Coastal Plain Research Station at Rocky Mount, NC. The rates of Dicamba to soybeans applied were 0.0, 1.08, 4.35, and 17.40 g ae ha-1. Evaluations were conducted at 14, 21, 28, 35 and 42 days after treatment (DAT). Results provides evidence that there no was difference of injury between the soybean maturity groups and varieties at all plantings and application timings.

INVESTIGATING POTENTIAL ANTAGONISM BETWEEN SEQUENTIAL APPLICATIONS OF DICAMBA AND GLUFOSINATE. E.A. Jones*, W.J. Everman, J.T. Sanders, D.J. Contreras, M. Granadino, and M. Fajardo, North Carolina State University, Raleigh, NC

ABSTRACT

Palmer amaranth has evolved resistance to seven herbicide groups and multiple herbicide-resistant populations are pervasive throughout the Southeast. The evolution of resistance to dicamba and glufosinate has not been confirmed in Palmer amaranth. Thus, farmers may begin to apply these herbicides in mixtures or sequentially to control multiple herbicide-resistant Palmer amaranth populations and mitigate the evolution of resistance to either herbicide. The mechanism of action is different for each herbicide. Dicamba is a slow-acting and systemic herbicide, while glufosinate is fast-acting and contact herbicide. Thus, the interaction between the two herbicide herbicides could result in control failures when tank-mixed or sequentially applied. Furthermore, the herbicides are height dependent, meaning a larger weed could exacerbate the control failure.

An experiment was conducted to determine the control of Palmer amaranth with tank mixes and various sequential applications of dicamba and glufosinate. The herbicide programs were applied to plants 5 and 15 cm in height. Sequential applications were made one week after the initial application. Treated Palmer amaranth plants were assessed weekly for percent weed control and height reduction for 5 weeks. Colby's equation was used to assess if tank mixes of dicamba and glufosinate at both weed sizes resulted in antagonism.

Colby's equation indicated no antagonism of tank-mixed dicamba and glufosinate at either weed size. The tank-mixed dicamba and glufosinate application at the 5 cm weed size provided 87% control compared to the application on the 15 cm weed size which achieved 62% control. The sequential applications of dicamba and glufosinate at both weed sizes provided the same control as the dicamba+glufosinate tank mix except for glufosinate fb dicamba (5 cm) and glufosinate fb glufosinate (15 cm). The reduction in growth was comparable to the visual efficacy evaluations for each herbicide program (R = 0.62).

The results of the experiment provide evidence that tank mixes and sequential applications of both herbicides are efficacious on Palmer amaranth of both sizes tested. The tested applications can be recommended to farmers for standard weed control or for controlling Palmer amaranth escaping an initial herbicide application. Since there is no difference in control between any combinations of sequential applications both herbicides should be included to reduced selection pressure.

EXPLORING *BRASSICA CARINATA* (A.) BRAUN TOLERANCE TO PREEMERGENCE AND POSTEMERGENCE HERBICIDES. S.A. Ramsey*, A. Post, T.A. Reinhardt Piskackova, M.E. Camacho, and R.G. Leon, North Carolina State University, Raleigh, NC

ABSTRACT

Brassica carinata (A.) Braun, or carinata, is an oilseed crop that is currently being developed for biofuel production. Southeastern growers have been interested in growing carinata as a winter crop because of biofuel industry demand and potential use as a rotational crop. As a new crop, there are no herbicides registered for use in carinata, so preliminary screening was used to identify herbicides safe for the use in carinata. This objective of this study was to assess the safety of select preemergence and postemergence herbicides at varying rates on carinata seedling establishment and plant growth. The preemergence herbicides used were diuron, napropamide, and clomazone. The postemergence herbicides used were simazine and clopyralid. The rates that were tested included a recommended label rate, one-quarter, one-half, two times, four times, and eight times the recommended label rate for each herbicide. Simazine and clopyralid were candidate postemergence herbicides for carinata production, while diuron could be used for control of volunteer carinata in rotational crops. In the future, further investigation of the development of cultural and mechanical approaches are needed to provide growers with more weed control options for carinata production.

SUSPECTED GLYPHOSATE RESISTANT COMMON LAMBSQUARTERS (*CHENOPODIUM ALBUM*) IN THE MID-ATLANTIC.
S.J. Michael*, M.L. Flessner, and S.D. Askew, Virginia Tech, Blacksburg, VA

ABSTRACT

Glyphosate has been at the forefront of nearly all herbicide programs since the introduction of glyphosate-tolerant crops in 1996. Since then, cases of glyphosate resistant weeds have increased and now, Common lambsquarters (*Chenopodium album*), could potentially be the next weed on that list in the Mid-Atlantic, US.

Common lambsquarters seeds from a suspected resistant population were obtained in Salisbury, MD from a farm that was having difficulty controlling common lambsquarters with glyphosate. The suspected resistant seeds were brought to Blacksburg, VA for greenhouse studies to confirm resistance. Seeds of a known susceptible population were obtained in Blacksburg, VA and used for a comparison population. Both populations were subject to sulfuric acid scarification to promote germination and were placed in separate trays in a greenhouse under red light. After germination, individual plants were transplanted into pots and were treated when plants were 5-10 cm tall. A dose response study with a randomized complete block design was used with the susceptible population having five replications and the putative resistant population having six replications. Treatments were glyphosate at 0, 0.001, 0.005, 0.01, 0.05, 0.1, 0.5, 1.0, 5.0, and 10 kg ha^{-1}. Visible injury ratings and above ground biomass were taken approximately four weeks after application. Data were subject to nonlinear regression using a three-parameter log-logistic curve. Subsequently, parameter estimates were used to generate I50 and I90 values (the glyphosate rate resulting in 50 or 90% control, respectively) for each population and determine a resistance factor.

An additional field study was conducted in Salisbury, MD where suspected resistant common lambsquarters seeds were collected. Herbicide applications were subject to different timings with atrazine (1.12 kg ha^{-1}) applied pre-emergence, atrazine, thifensulfuron-methyl (0.04 kg ha^{-1}), glyphosate (1.12 kg ae ha^{-1}), dicamba (0.56 kg ae ha^{-1}), and glyphosate+dicamba applied post-emergence when common lambsquarters were 5-10 cm tall, and glyphosate, dicamba, and glyphosate+dicamba applied when common lambsquarters were 10-20 cm tall. This study was randomized using complete block design with three replications. Visible injury ratings were taken 7 days after application B and 7 days after application C (10-20 cm tall common lambsquarters). Data was subjected to ANOVA followed by means separation using Fisher's protected $LSD_{0.05}$.

Dose response curves from greenhouse studies indicated that a 1.46 kg ha^{-1} increase in dose is required for 90% control of the suspected resistant population compared to the susceptible population and a 0.85 kg ha^{-1} increase in dose to control 50% of the suspected resistant population compared to the susceptible population. Field study results showed resistance to atrazine and decreased effectiveness of glyphosate relative to treatments of thifensulfuron-methyl, dicamba, and glyphosate+dicamba.

Dose response studies have shown a reduced impact of glyphosate on common lambsquarters in the suspected resistant population. Future studies will investigate the mechanism of resistance.

WEED CONTROL AND CROP TOLERANCE WITH PREEMERGENCE HERBICIDES IN SWEET BASIL. M. Wasacz*, A. Wyenandt, and T.E. Besancon, Rutgers University, New Brunswick, NJ

ABSTRACT

The number of herbicides labeled for use on sweet basil remains extremely limited with only seven active ingredients. Therefore, many herb growers use plasticulture and mulching as a method of weed control, which can be ineffective on many weed species, including yellow nutsedge (*Cyperus esculentus* L.). Consequently, there is a need for safe and effective herbicides in sweet basil. Sulfentrazone (140, 210, and 280 g a.i. ha^{-1}), ethalfluralin (840 g a.i. ha^{-1}), linuron (560 g a.i. ha^{-1}), fomesafen (150 g a.i. ha^{-1}), bensulide (5.6 kg a.i. ha^{-1}), pronamide (1.12 kg a.i. ha^{-1}), and diuron (1.12 kg a.i. ha^{-1}) applied PRE at seeding were evaluated in a field study in Bridgeton, NJ. Napropamide is the only labeled PRE herbicide on sweet basil and was included as a standard (2.24 kg a.i. ha^{-1}). Ratings included crop injury (stunting, chlorosis, and necrosis), crop stand counts, fresh yield, and weed control. Use of ethalfluralin, bensulide, or pronamide drastically reduced fresh yield, on average, by 80%, whereas plots sprayed with diuron had no harvestable plants. Diuron, fomesafen, and sulfentrazone provided excellent control ($\geq$ 96%) of smooth pigweed (*Amaranthus hybridus* L.) 8 WAT, while goosegrass [*Eleusine indica* (L.) Gaertn.] control was excellent ($\geq$ 98%) in plots treated with napropamide, diuron, and sulfentrazone at 210 or 280 g a.i. ha^{-1}. Of all herbicides tested, sulfentrazone at 210 g a.i. ha^{-1} and linuron at 560 g a.i. ha^{-1} caused the least amount of crop injury while providing optimal weed control, resulting in the highest sweet basil yield with more 18,000 kg fresh biomass ha^{-1}.

FALL AND SPRING HERBICIDE PROGRAMS FOR MARESTAIL MANAGEMENT IN CEREAL RYE. J.S. Adam*, J.M. Wallace, and D. Lingenfelter, Penn State University, University Park, PA

ABSTRACT

Fall and early spring horseweed (*Erigeron canadensis* L.) management is increasingly recommended in no-till systems within the Northeast region. Herbicide applications and cover crops can both provide horseweed suppression in the fall and spring, but few guidelines are available for integrating these control tactics. In a preliminary experiment from 2018 to 2019, fall- and spring- applied herbicides were evaluated in a fall-sown cereal rye (*Secale cereale* L.) cover crop to quantify differences in glyphosate-resistant horseweed control levels and cereal rye biomass production at Penn State's Russell E. Larson Agricultural Research Center in Rock Springs, PA. Cereal rye was sown at 67 kg ha^{-1} on September 27 and fall and spring herbicides were applied on October 19 and April 10, respectively. Applications were made with a hand-held boom that delivered 140 L ha^{-1}. Herbicides included 2,4-D ester at 560 and 1120 g ai ha^{-1}, dicamba at 280 g, sulfentrazone plus metribuzin premix at 379 g, and chlorimuron plus tribenuron premix at 46 g. The premix treatments were tank-mixed with 2,4-D ester (560 g). At the fall application timing, horseweed ranged from cotyledons to 1.5 cm in rosette diameter and cereal rye was 8 to 10 cm tall (1-2 tillers). At the spring application timing, horseweed rosette diameter ranged from 1.5 to 5 cm and cereal rye was 8 to 15 cm tall (7-12 tillers). Prior to soybean planting (May 15), rye biomass and horseweed density were quantified. Cereal rye biomass in the untreated check was 3,175 kg ha^{-1} at termination. Herbicide treatments did not affect cereal rye biomass production in comparison to the untreated check, with the exception of sulfentrazone + metribuzin, which resulted in lower cereal rye biomass in the fall (2,387 kg ha^{-1}) and spring (2,266 kg ha^{-1}) treatments. Relative to the untreated check, horseweed density was significantly reduced (93 to 98%) in all herbicide treatments except for spring applications of 2,4-D applied alone at the low rate (560 g ai ha^{-1}) and in combination with chlorimuron + tribenuron, which resulted in 68 and 79% horseweed population reductions, respectively. Results of our first experimental year suggest that integrating a rye cover crop with fall- or spring- applied herbicides can be an effective tool to control horseweed in no-till systems. Our results also suggest that 2,4-D applied at a low rate (560 g ai ha^{-1}) in the spring may not be an effective option for horseweed control regardless of additional suppression from cereal rye competition. Sulfentrazone + metribuzin reduced rye biomass following fall and spring application timings, which may limit horseweed suppression benefits from surface rye mulch following termination.

UPDATE ON 2019 WEED SCIENCE RESEARCH IN THE IR-4 ENVIRONMENTAL HORTICULTURE PROGRAM AND 2020/2021 PRIORITIES. C.L. Palmer* and E. Vea, IR-4 @ Rutgers University, Princeton, NJ

ABSTRACT

In 2019, the IR-4 Environmental Horticulture Research Program sponsored research on several weed science projects: pre- and postemergent herbicide crop safety with over-the-top in-season applications, efficacy of post-emergent herbicides, and Christmas tree crop safety with herbicides used to manage cover crops. For the over-the-top herbicide crop safety screening the goal was to develop lists of crops where applications would not harm woody and herbaceous perennials grown in container nurseries. Products tested included Basagran (bentazon), Biathlon (oxyfluorfen + prodiamine), Dimension 2EW (dithiopyr), Dismiss 4F (sulfentrazone), F6875 4SC (sulfentrazone + prodiamine), Fiesta Herbicide (FeHDTA), Fortress (isoxaben + dithiopyr), Freehand G (pendimethalin + dimethenamid-p), Gallery SC (isoxaben), Gemini Granular (prodiamine + isoxaben), Marengo 74SC (indazaflam), Pendulum G (pendimethalin), Ronstar G (oxadiazon), SP1770, and Tower EC (dimethenamid-p). Applications were made at dormancy (preemergence) or when leaves were fully expanded (postemergence) and then approximately 6 weeks later for all products. Basagran was screened on 19, Biathlon on 7; Dimension EW on 6; Dismiss on 3; Fiesta on 30; Fortress on 13; Freehand on 2; Gallery on 1; Gemini Granular on 4; Marengo on 8; Pendulum G on 6; SP1770 on 8; Ronstar G on 3; and Tower was screened on 4 crops. The results from this research will aid in the development of product labels and will help growers and landscape care professionals make more informed product choices.

POPULATION STRUCTURE AND DIVERSITY OF PARASITIC DODDER IN MASSACHUSETTS CRANBERRY BOGS. J. Scott, H.A. Sandler*, K. Ghantous, and A. Caicedo, UMass Cranberry Station, East Wareham, MA

ABSTRACT

Dodder (*Cuscuta* spp.) is the only parasitic plant known to occur on cranberry farms and its weedy infestations can result in 50-100% yield loss. After germination, dodder seedlings attach and penetrate cranberry plant stems, and extract water and nutrients from the host. In a recent survey, Massachusetts (MA) cranberry growers reported that 30-67% of their acreage was infested with dodder, with 7% and 52% of growers listing dodder as a severe or moderate problem. Its prolific seed production and long-lived seedbank (>20 yr) make dodder one of the most damaging cranberry weed pests in MA. However, a hurdle to implementing optimal dodder management is our lack of knowledge about the various biotypes found in cultivated cranberry bogs. To date we do not know the number or identity of dodder species infesting cultivated bogs, the extent of genetic diversity present in these infestations, nor the source of dodder plants colonizing commercial bogs. In summer of 2018 we collected 426 dodder samples from 32 commercial bogs and six non-agricultural environments in MA. We observed variation in stem diameter and stem color across collected samples. We also obtained 80 dodder samples from New Jersey (8 sites, 3 were native wetlands), and 16 samples from Wisconsin (8 commercial sites). We have developed an in-house genotyping-by-sequencing (GBS) protocol to carry out genome-wide characterization of single nucleotide polymorphism (SNPs) in dodder. Generation of this high-resolution genetic data is underway, and will allow us to answer the following questions about dodder infesting cranberries: 1) Which dodder species are the ones that occur in commercial cranberry bogs? 2) How much genetic diversity is present in dodder infestations, and how is this diversity structured across different sites? 3) What is the relationship between weedy dodders in cranberry bogs and wild dodders that occur outside of these commercial sites?

THE ENDLESS PURSUIT OF PRODUCTIVITY, POTENCY, AND SELECTIVITY: EXOTIC SHRUB EDITION. A.E. Gover*, A.C. Causey, and M.L. Fuchs, Penn State, University Park, PA

ABSTRACT

[no abstract entered]

ECOLOGICAL AND EVOLUTIONARY DYNAMICS AT THE RANGE EDGE OF AN EXPANDING INVASIVE SPECIES. R. Fletcher, D. Atwater, D. Haak, and J.N. Barney*, Virginia Tech, Blacksburg, VA

ABSTRACT

When an exotic invasive species successfully colonizes its new range, rapid evolution is thought to play an important role in allowing the invader to overcome the strong selection pressures imposed on it by the novel environmental conditions of its new range. At the continental-scale, local adaption of exotic species has commonly been observed in response to climatic variation however, less is known at the finer, landscape-scale. There is some evidence that exotic species may adapt in response to landscape heterogeneity (e.g., Scott et al. 2010; Godoy et al. 2011). A study has found phenotypic and genetic differentiation in agricultural and non-agricultural populations of *Sorghum halepense* (Johnsongrass), a global invasive species and agricultural weed. In this study, *S. halepense* was used to determine whether the phenotypic differentiation of agricultural and non-agricultural populations is the result of fine-scale, local specialization. We found that populations outside the core range, which we call periphery, we less than half the size of core populations, but flowered much earlier. However, despite the smaller size still produced the same number of seeds. This suggests that there is a trade-off between flowering time and investment in growth. Our results suggest that Johnsongrass has the adaptive and reproductive potential to continue expanding its range north, though this expansion may be limited by other abiotic factors, such as cold temperature or drought. As climate changes understanding invasive species ability to adapt, or not, will inform future range expansion and the eco-evolutionary responses of these introduced species.

INVASIVE PLANTS REDUCE THE ABUNDANCE OF NATIVE ANIMALS, ESPECIALLY IN WET ECOSYSTEMS. R. Fletcher, R.K. Brooks, V.T. Lakoba, G. Sharma, A. Heminger, C. Dickinson, and J.N. Barney*, Virginia Tech, Blacksburg, VA

ABSTRACT

Despite our growing understanding of the impacts of invasive plants on native plants, important gaps remain. Importantly, we still do not understand whether native and exotic species respond differently to plant invasion following introduction. Furthermore, we know relatively less about the effects of invasive plants on resident animals. Therefore, we performed a meta-analytic review of the effects of invasive plants on native and exotic resident animals: 70 studies were identified for the review, of these 12 studies reporting 206 observations qualified for inclusion in the meta-analysis. We found that invasive plants reduced the abundance of native animals, but not exotic animals. This effect varied by animal phyla, with invasive plants reducing the abundance of native annelids and chordates, but not molluscs or arthropods. We found that animals in 'wet' and 'dry' ecosystems were differently affected, but found no differences among animal trophic levels (e.g., herbivore, predator). Additionally, the impact of invasive plants increased over time, but this did not vary with animal nativity. Our review found that few studies did not identify animals to species, severely limiting our ability to understand the impacts of invasive plants. We provide an important first insight into how native and exotic species respond differently to invasion, the consequences of which may facilitate cascading trophic disruptions. We call for more rigorous studies of invaded community impacts across taxa, and most importantly, explicit consideration of resident biogeographic origin.

NEW YORK RESOURCES FOR WEED ID. C. Marschner*, S. Bachman, and A. DiTommaso, Cornell University, Ithaca, NY, NY

ABSTRACT

The Weed Ecology and Management Lab at Cornell University is working with extension educators and specialists and certified crop advisors to expand the weed identification resources available for New York agriculture. In 2019, we constructed a website for the program; in 2020 we will be conducting weed identification workshops around the state, and constructing New York specific materials for difficult-to-identify weed complexes (e.g. pigweeds) and locations of problem weeds in New York, answering weed identification questions, and providing updates on weed ID through the growing season.

HOW WELL DO SMARTPHONE PLANT ID APPS WORK? A WEED SCIENTIST TAKES A LOOK AT THE TECHNOLOGY. L. Sosnoskie*, Cornell University, Geneva, NY

ABSTRACT

The U.S. loves smartphones; the statistics portal Statista.com estimates 81% of all U.S. adults will own at least one device in 2019. Mobile applications (apps) account for over 90% of users' smartphone internet time, with most of it focused on socializing, gaming and other forms of entertainment. Growers, crop consultants and members of allied industry have asked if there are any available smartphone apps that can scan/evaluate/assess pictures of weeds and return a proper identification to you? As it turns out, there are. One popular tool is the Pl@ntNet app, which allows users to upload one or more pictures of leaves, flowers, fruits or bark and then compares the photos to a database (there are multiple image galleries from which a user can screen against) using visual recognition software. The app then returns a list of possible matches (hits) and provides the user with links to the Wikipedia entries for each of the species (in case more information is desired). Another app is iNaturalist. Like Pl@ntNet, the iNaturalist app compares a user's picture against those in an image bank and provides users with a list of suggestions, links to information about the species and maps of occurrences worldwide. Both Pl@ntNet and iNaturalist allow you to share your images with their communities. For more information about the history of the apps, databases, curation, citizen science, and more see: https://plantnet.org/en/ and https://www.inaturalist.org/.

Results from two studies to evaluate app performance for weed identification suggest that the age of the plant, the tissue selected for use in the process, and, likely, the genus and family (not testable due to small numbers of samples) of the specimen influence success. The quality of the images (particularly the amount of background that can be a distraction) is also another factor that affects the ability to achieve an ID. The taxonomic knowledge of the person making the submission may also impact app performance; a trained weed scientist may bias results, favorably, by purposely selecting and submitting images highlighting characteristic/distinguishing traits. I recommend my clientele always double-check ID's against a trusted source (e.g. guidebooks, knowledgeable extension staff, curated/vetted web sites).

Selected methodology and results include:
Study 1. Evaluate the performance of Pl@ntNet for identifying weeds in California. Pl@ntNet database selected: Plants of the United States. 60 images of 46 species (39 broadleaf, 7 grass/grass-like) in 21 plant families were submitted to the app. Most represented families: Asteraceae (14), Poaceae (7), Malvaceae (6), Amaranthaceae (4). All images were of adult (non-seedling) plant specimens growing under natural/field conditions. Image breakdown: 31 of leaves, 8 of leaves/flowers, 14 of flowers, 7 of flowers/fruit. Specimen identity confirmed using 'Weeds of California and Other Western States' (ISBN-13: 978-1879906693). 72% of specimens were identified correctly by 1st (60%) and 2nd (12%) hits in Pl@ntNet. 15% of specimens did not have a correct identification (no hits) recommended by Pl@ntNet. 86% of floral images and 75% of leaf/flower images were identified correctly (1st hit) by Pl@ntNet. 52% of leaf images and 43% of flower/fruit images were identified correctly (1st hit) by Pl@ntNet.

Study 2. Compare Pl@ntNet to iNaturalist for identifying weeds in California and New York. Pl@ntNet database selected: Plants of the World (to match iNaturalist's worldwide database (not selectable)). 60 images of 60 species (45 broadleaf, 15 grass/grass-like) in 23 plant families were submitted to the apps. Most represented families: Poaceae (13), Asteraceae (13), Amaranthaceae (5). Seedlings (12) and adult (48) plants were both represented in the images. All specimens were growing under natural/field conditions. Image breakdown: 27 of leaves, 5 of leaves/flowers, 24 of flowers, 4 of flowers/fruit. Specimen identity confirmed using 'Weeds of California and Other Western States' (ISBN-13: 978-1879906693). For broadleaves, 78% of specimens were identified correctly by 1st hit in iNaturalist. For broadleaves, 69% of specimens were identified correctly by 1st hit in Pl@ntNet. For broadleaves, 71% of Naturalist's and Pl@ntNet's 1st hits matched each other. iNaturalist and Pl@ntNet correctly identified adult broadleaves 85% and 73% of time, respectively, with the 1st hit (58% correct for seedlings). 89% of floral images (both apps) and 100% (iNaturalist) and 75% (Pl@ntNet) of flower/fruit images were identified correctly with 1st hit. 67% (iNaturalist) and 63% (Pl@ntNet) of leaf images and 100% (iNaturalist) and 60% (Pl@ntNet) of leaf/flower images were identified correctly with 1st hit. All images of grass/grass-like specimens were of flowers/flower spikes/flower clusters. For grasses, 47% of specimens were identified correctly by 1st hit in iNaturalist, 33% of specimens were identified correctly by 1st hit in Pl@ntNet. For grasses, 33% of Naturalist's and Pl@ntNet's 1st hits matched each other.

WEED SEED RAIN PHENOLOGY OF ECONOMICALLY IMPORTANT WEEDS IN THE MID-ATLANTIC, NORTH- AND SOUTH-CENTRAL REGIONS OF THE U.S. L.S. Shergill*, L. Lazaro, J. Evans, M.J. VanGessel, J. Norsworthy, A. Davis, K. Bradley, L.E. Steckel, M. Bagavathiannan, J. Bond, J.L. Lindquist, N. Jordan, M.L. Flessner, W.J. Everman, N. Korres, and S. Mirsky, USDA-ARS & University of Delaware, Beltsville, MD

ABSTRACT

Multiple herbicide-resistant (MHR) weeds are challenging sustainable crop production as herbicides are rapidly becoming less effective and herbicide discovery has slowed. New integrated weed management (IWM) practices are urgently needed that rely on multiple weed control strategies. One promising tactic for managing MHR weeds is Harvest-time Weed Seed Control (HWSC), in which weed seeds are removed/destroyed at harvest time to reduce soil seedbank. The primary factor on which success of HWSC practice relies is the biological attribute of seed retention at crop maturity enabling its collection and processing at crop harvest. Therefore, field experiments were conducted at multiple sites across 14 states in mid-Atlantic, north- and south-central regions of the United States to quantify the phenology of weed seed rain in soybean. Majority of broadleaf weeds shatter <10% seed at soybean maturity. Palmer amaranth (AMAPA), smooth pigweed (AMACH), hemp sesbania (SEBEX), lambsquarters (CHEAL), jimsonweed (DATST), & common cocklebur (XANST) retain >90% seed until 3 wks after soybean maturity across majority of sites. Common waterhemp (AMATA) seed retention was variable across sites, with seed shatter <10% until 2 wks after soybean maturity at 50% of the sites. Johnsongrass (SORHA) retained >90% seed up to 3 wks after soybean maturity across all site-years. Barnyardgrass (ECHCG) & jungle rice (ECHCO) seed retention was >80% until 2 wks after soy maturity. Overall, broadleaf weeds tend to retain more seeds overtime compared to grasses. Many annual weed species retain high proportion of seeds on the plants at crop maturity as well as following delays in crop harvest. Results indicate that HWSC practices have good potential for several weed species in the US.

GETTING RID OF WEEDS THROUGH INTEGRATED WEED MANAGEMENT. C.G. Rubione*, M.J. VanGessel, S. Mirsky, K.B. Pittman, L. Lazaro, M. Bagavathiannan, M.L. Flessner, L.S. Shergill, and V. Ackroyd, University of Delaware, Salisbury, DE

ABSTRACT

Integrated weed management (IWM) combines various methods to reduce or eliminate the effect of weeds on crop production over time. These weed management methods form a "toolbox" in which "tools" can be integrated into a weed management plan catered to the particular farm and problem. The toolbox includes preventative, biological, chemical, cultural, and mechanical strategies. IWM also considers the weed species present and tailors strategies to these species.

In conventional crops, integrated weed management is not a replacement for herbicides. For many decades, herbicides have been the primary means of weed management due to their simplicity, effectiveness, and affordability. However, relying too much on a few herbicides has led to an increase of weed species that are not effectively controlled with the herbicide program or selecting for herbicide-resistant biotypes. IWM approaches go beyond relying on herbicide rotation and mixtures. IWM programs use all available methods that will best solve the problem.

To expand the use of IWM, GROW (Get Rid Of Weeds) was developed as part of an USDA-ARS Areawide Project. GROW is a place to find helpful, research-based resources for IWM of herbicide-resistant and -susceptible weeds. GROW uses the website www.growiwm.org as a main source of information, as well as social media accounts (Twitter, Instagram, YouTube, and Facebook).

The website provides a resource for general interest content on IWM, current and ongoing research projects, and more in-depth information on specific management strategies. When navigating GROW's website users can either browse for general information such as "What is IWM? or "A-B-C's of IWM Principles" or go into more specific areas such as "IWM Toolbox" or "Resistant Weeds".

The website is currently undergoing a new design to improve the user experience. GROW is an opportunity for anyone working on IWM to have a platform to expand their reach and always looking for new partners.

INVESTIGATING THE LONG-TERM EFFECTS OF COVER CROP ROTATIONS AND REDUCED TILLAGE. G.A. Bagley*, S. Mirsky, and B.W. Davis, North Carolina State University, Beltsville, MD

ABSTRACT

The Cover Crop Systems Project (CCSP) is a Long Term Agricultural Research (LTAR) site located in Beltsville, MD that has been established for 4 field seasons. It is designed to investigate the effects of cover crop biomass and reduced tillage on metrics such as cash crop yield and weed biomass in two organic and four conventionally managed cropping systems. All systems are in a three year corn-soy-wheat rotation; the four conventional systems are no-till with double cropped beans and a gradient of cover crop inclusion, and the organic systems vary in their use of tillage.

MULTIPLE HERBICIDE RESISTANCE IN *LOLIUM* SPP. V. Singh*, J. Swart, and M. Bagavathiannan, Virginia Tech, Painter, VA

ABSTRACT

Field surveys were conducted across the Blacklands region of Texas to document the distribution of herbicide resistant *Lolium* spp. infesting winter wheat production fields in the region. A total of 68 populations (64 Italian ryegrass and 4 perennial ryegrass) were evaluated in a greenhouse in 2018 for sensitivity to herbicides of three different sites of action, an ALS-inhibitor (mesosulfuron-methyl), two ACCase-inhibitors (diclofop-methyl and pinoxaden), and an EPSPS-inhibitor (glyphosate). Herbicides were applied at 2X the label recommended rates for mesosulfuron-methyl (29 g ai ha^{-1}), diclofop-methyl (750 g ai ha^{-1}), and pinoxaden (118 g ai ha^{-1}); and 1X rate for glyphosate (868 g ae ha^{-1}). The herbicide screenings were followed by dose-response assays of the most-resistant ryegrass population for each herbicide at eight rates (0.5, 1, 2, 4, 8, 16, 32, and 64X), compared to a susceptible population at six rates (0.0625, 0.125, 0.25, 0.5, 1, and 2X). The initial screening, as well as dose-response experiments, were conducted in a completely randomized design with three replications and two experimental runs. Survivors (<80% injury) were characterized as highly resistant (0-20% injury) or moderately resistant (21-79%). Results showed that 97, 92, 39, and 3% of the Italian ryegrass populations had survivors to diclofop-methyl, mesosulfuron-methyl, pinoxaden, and glyphosate, respectively. Of the four perennial ryegrass populations, three were resistant to diclofop-methyl and mesosulfuron-methyl, and one of them was resistant to pinoxaden as well. Perennial ryegrass populations were not resistant to glyphosate. Dose-response assays revealed 40-, 199- and 23-fold resistance in Italian ryegrass to mesosulfuron-methyl, diclofop-methyl, and pinoxaden, respectively compared to a susceptible standard. One Italian ryegrass population exhibited three-way multiple resistance to ACCase-, ALS-, and EPSPS-inhibitors. The proliferation of multiple herbicide resistant ryegrass is a challenge to sustainable wheat production in Texas Blacklands and warrants diversified management strategies.

UNMANNED AERIAL SYSTEM BASED WEED DETECTION AND HERBICIDE SPRAY APPLICATIONS. V. Singh*, D. Martin, B. Sapkota, M. Bishop, and M. Bagavathiannan, Virginia Tech, Painter, VA

ABSTRACT

Weed control using unmanned aerial vehicles (UAVs) has emerged as an idea of the new age digital agriculture. UAV-based aerial herbicide applications are beneficial even to small landholders as it is expected to be effective across scales. Field experiments were conducted in 2019 to evaluate the efficacy of herbicides in controlling prominent weed species in Liberty Link® Soybean with UAV based spray and compared with conventional backpack spray applications. Grass and broadleaf weed seeds were broadcasted and weed density was recorded before herbicide spray application. UAV based spray applications were conducted at 18.7 and 37.4 L ha^{-1} and backpack spray applications were conducted at 140 L ha^{-1}. Droplet spectra was measured using water sensitive papers (WSP) strips collectors. The test was repeated in time, each with four replications. Plant mortality was scored on a scale of 0 to 100%, where 0 = no mortality, 100 = 100% mortality. DayGlo fluorescent dyes were used separately to quantify spray droplets on weed leaves using digital imaging techniques. Soon after the spray application, the spray droplet captured on plant leaves were analyzed by the DropletScan™ scanner-based software system. The droplet size spectra parameters were examined $D_{v0.1}$, $D_{v0.5}$, $D_{v0.9}$, droplet density (droplets/cm^2), percent area coverage and percent applied collected on WSP collectors. For herbicide spray study, weed control, and crop injury were recorded at 15 and 21 days after herbicide treatment. UAV spray treatments resulted in higher droplet density at abaxial leaf surface (25% of total) which may increase the efficacy of the contact herbicides. Herbicide treatments with UAV provided >90% weed control which was greater than backpack spray application when applied late POST.

WINTER ANNUAL WEED SUPPRESSION FROM COVER CROP MONOCULTURES AND MIXTURES AND FLUMIOXAZIN. L.S. Rector, M.L. Flessner*, K.B. Pittman, D.S. McCall, and W.E. Thomason, Virginia Tech, Blacksburg, VA

ABSTRACT

Winter weeds can interfere with cash crop planting. Cover crops are known to compete with winter weeds, but little research has evaluated multispecies cover crops. Field studies were conducted in Blacksburg, Virginia in 2017 and 2018, Christiansburg, Virginia in 2017, and West Point, Virginia in 2018 to determine the ability of early planted cover crops to suppress winter weeds prior to cash crop planting and compare that to flumioxazin applied at cover crop planting. Cover crops were planted in September, fertilized, and terminated in early May to maximize cover crop biomass. Cover crop treatments consisted of monocultures, two-way, and three-way mixtures of cereal rye, crimson clover, hairy vetch, and forage radish. Cover crop mixtures produced greater biomass compared to monocultures as a group, due to less biomass of the legume monocultures. Flumioxazin provided similar control as the average of all cover crops across all weed species rated in December. At cover crop termination, flumioxazin provided less Persian speedwell control and similar control of purple deadnettle and yellow woodsorrel compared to the average of all cover crops. Cereal rye containing treatments provided >85% control of all weed species at cover crop termination. As cover crop biomass increased to 6000 kg ha^{-1}, winter weed biomass decreased at cover crop termination to less than 200 kg ha^{-1}. Cereal rye containing treatments were more effective at reducing winter annual weed density in mid-March compared to legume monocultures and forage radish. No difference was observed in winter annual weed control between cereal rye alone or in a mixture. Cereal rye can be used in a mixture for goals of receiving additional agronomic benefits of a legume or brassica cover crop species without compromising winter annual weed control. Cereal rye is more effective at controlling winter weeds prior to cash crop planting than flumioxazin.

IMPACT OF RUMINANT DIGESTION ON GERMINATION OF INGESTED SEEDS OF NATIVE WARM-SEASON GRASSES AND AGRONOMIC WEED SPECIES IN THE SOUTHEAST UNITED STATES. L.U. Snyder* and W.J. Everman, North Carolina State University, Zebulon, NC

ABSTRACT

Dissemination of seeds by livestock has been discussed as a means of spreading both desired and problematic plant species, across grasslands, pastures and agronomic areas through various means. Excretion of seeds and spread of manure as a fertility amendment can account for colonization of new plants. The dispersal of high nutritional-quality native grasses is of value to grassland conservation and restoration, and to livestock producers as use for forage. Likewise, problematic weed species can be spread through means of livestock fertility amendments and supplemental feed. Seeds that retain viability subsequent to ruminant digestive tract exposure can be vital to the process of introducing new desirable forage species and enhancing the spread of established species. The objective of this study was to examine the effects of the rumen and intestinal digestion on seed germination of native warm-season grasses and particularly problematic agronomic weed species in the Southeast U.S. affecting crop yield. Two experiments were conducted to determine how ruminant digestion influences seed viability and germination on native warm-season grasses for potential forage, biomass and restoration use and on problematic weeds in agronomic settings. Experiment one was to determine the impact of rumen exposure and digestion on seed viability and germination. Seeds were treated to in situ (live animal) rumen digestion in a cannulated steer for 48 hours. Experiment two was to examine the effect of small intestine exposure and digestion to determine the impact on seed viability and germination and potential points of digestion which cause or impact seed mortality. Seeds were treated to in vitro small intestine digestion for 24 hours. Seed germination varied by plant species and response to respective treatment. The combination of rumen phase and small intestine procedure (SIP) exposure (abomasal-simulated digestion) resulted in the highest seed mortality. Native-warm season grass seed, such as switchgrass (Panicum virgatum L.) and big bluestem (*Andropogon gerardii* L.), retained viability but germination after digestion exposure was reduced more than agronomic weed species. Switchgrass mean germination percentage was reduced by 86.70% and big bluestem mean germination percentage was reduced by 79.59% with exposure to rumen phase plus SIP. Troublesome agronomic weed seed, such as Palmer amaranth (*Amaranthus palmeri* L.) and Texas panicum (*Panicum texanum* L.), retained viability but germination declined with digestion exposure. Palmer amaranth mean germination percentage was reduced by 53.77% and Texas panicum mean germination percentage was reduced by 56.83% with exposure to rumen phase plus SIP, which was the highest rate of germination. Herbivory is a key process in range and grassland ecosystems and it is imperative to consider the role of herbivores as part of a holistic system and utilize this understanding to improve management approaches. Further research is needed in this area to determine and to understand how grazing animals, including cattle and other species, can impact the dissemination of seeds and establishment of plant populations.

SILAGE CORN RESPONSE TO BURCUCUMBER (*SICYOS ANGULATUS*) COMPETITION AND DROUGHT IN NEW YORK. K.M. Averill, A. DiTommaso*, S.H. Morris, and M.C. Hunter, Cornell University, Ithaca, NY

ABSTRACT

Climate change can alter weed population growth and spread and, because extended droughts in particular are expected to affect the Northeastern United States, more information is needed about how weed species respond to altered environmental conditions. Burcucumber is a challenge to manage in the Southeastern and the Mid-Atlantic US. This weedy annual vine is increasing in NY and poses a threat to agronomic crop production. Our work aimed to increase understanding of the potential influence of extended drought during the growing season on burcucumber competition in glyphosate-tolerant corn (*Zea mays* L.) silage production. Thus, in Ithaca, NY in 2018 and 2019, a competition and drought experiment was conducted to assess the effect of burcucumber on corn silage yield. Corn was planted in late May and burcucumber seedlings (of similar size to corn seedlings, both ~15 cm) were transplanted into corn rows in June, 10 days after planting (DAP) in 2018 and 35 DAP in 2019, at four densities (0, 0.5, 2, 3 plants m^{-2}). Other weeds were controlled with glyphosate, inter-row cultivation, and weekly hand-weeding. During early season corn growth, a 38-day drought in 2018 and a 21-day drought in 2019 were created using rainout shelters, 2.7 m wide by 3 m long by 3 m tall, made of steel electrical conduit, high clarity plastic, and gutters and drainage tubing. Soil water availability was estimated weekly during the experiment using fixed-in-place gypsum block electrodes. Corn silage and aboveground burcucumber biomass were harvested 108 DAP in 2018 and 112 DAP in 2019. Regression and mixed model ANOVA were conducted with drought, burcucumber density, year, and their interactions as fixed effects and block and corn density as random effects. Estimated soil water availability was lower in drought plots (47 ± 1 in 2018 and 56 ± 3 in 2019) compared to non-drought control plots (69 ± 1 in 2018 and 68 ± 1) ($P < 0.001$). Results showed that burcucumber competition ($P = 0.002$) and drought ($P = 0.02$) reduced fresh corn silage yield by 20% in the drought treatment plots (17 ± 1 Mg ha^{-1}) compared to the non-drought plots (21 ± 1 Mg ha^{-1}); the interaction between these factors was not significant ($P = 0.6$). Year and interactions with year were not significant ($P > 0.05$). Conducting the experiment on soil other than the fine sandy loam present in our study field or in a system other than continuous corn could produce different results. Overall, our work suggests that burcucumber infestation and extended drought during the growing season each have the capacity to reduce corn silage yield, but no evidence was found for an exacerbated effect on yield where the stressors were combined.

COMMON PURSLANE CONTROL AND CROP RESPONSE WITH PRE AND POST HERBICIDE APPLICATION IN SWEET PEPPER. T.E. Besancon, B.L. Carr*, M. Wasacz, and W. Kline, Rutgers University, Chatsworth, NJ

ABSTRACT

In 2018, New Jersey produced 48 million kg of peppers at a farm value of $40 million (USDA 2018). New Jersey pepper production is concentrated in the southern counties where well-drained loamy sand soils are optimal for this crop but also for common purslane (*Portulaca oleracea*), one of the most common and troublesome weed issues in NJ pepper fields. Common purslane can easily grow back from POST non-systemic herbicide applications, compete with neighboring pepper plants, and impede pepper harvest. Currently labeled herbicides effective at controlling common purslane and safe to the crop are limited to some PRE herbicides that will provide only a few weeks of residual control. Field studies were conducted in 2019 in Vineland, NJ, to evaluate weed control efficacy and pepper tolerance (variety 'Playmaker') to common purslane management programs based on PRE and/or POST herbicide applications between rows of plastic mulch. Treatments consisted of pendimethalin at 1.07 kg ai ha^{-1} or *S*-metolachlor at 1.07 kg ai ha^{-1} applied PRE alone or combined with fomesafen at 0.42 kg a.i. ha^{-1}, fomesafen PRE alone at 0.32 kg ai ha^{-1}. PRE application of pendimethalin / *S*-metolachlor plus fomesafen were followed or not by POST application of imazosulfuron at 0.21 kg ha^{-1} when emerged common purslane seedlings averaged 5-cm. height. POST application of imazosulfuron alone at 0.21 or 0.34 kg a.i. ha^{-1} was also included in addition to non-treated weedy and non-treated weed-free checks. Crop injury and weed control data were evaluated 28 and 56 DA-PRE, equivalent to 20 DA-POST. Peppers were harvested three times to determine yield weight accordingly to USDA standard grades. Fomesafen applied alone at 0.32 kg ai ha^{-1} or combined at 0.42 kg a.i. ha^{-1} with other PRE herbicides provided at 60% to 75% common purslane control 56 DA-PRE, compared to less than 30% for pendimethalin or *S*-metolachlor alone. Following *S*-metolachlor or pendimethalin applied PRE with Imazosulfuron POST allowed to maintain over 75% common purslane control by 56 DA-PRE. Treatments that included *S*-metolachlor applied PRE provided ≥ 95% large crabgrass (*Digitaria sanguinalis*) control 56 DA-PRE, regardless of tank-mix partner or POST application. Hairy galinsoga (*Galinsoga quadriradiata*) is another troublesome species in most vegetable crops that was controlled ≥ 90% 56 DA-PRE with fomesafen PRE or imazosulfuron POST, regardless of application rates or PRE treatments. No crop injury was observed throughout the duration of the trial. Yield weights did not significantly differ across treatments in comparison to the untreated weed-free check

WEED COMMUNITIES IN U.S. PROCESSING SNAP BEAN. C.J. Lowry*, N. Korres, J. Colquhoun, E. Peachey, and M. Williams, USDA, Champaign, IL

ABSTRACT

Snap bean (*Phaseolus vulgaris*), the vegetable form of dry bean, is an important vegetable crop with a value of $372 million. As with other specialty crops, herbicide options within snap bean production are limited because of crop sensitivity concerns. Currently, information is lacking on the scope of weed issues in snap bean grown for processing across the US, and how agronomic and environmental factors contribute to weed pressure. The objectives of our research were to determine the weed composition and management practices in snap bean grown for processing in the U.S. In Summer 2019, we conducted a survey of weed communities in snap bean fields in Illinois, Oregon, and Wisconsin. In Illinois, we surveyed 25 famers' fields in June and 17 fields in September. The weed species with the highest relative frequency (percent of farm sites in which species occurred) in the first snap bean crop included *Digitaria* spp., *Mollugo verticillata*, *Ipomoea hederacea*, *Solanum ptychanthum*, and *Amaranthus* species. The weed species with the highest relative frequency (percent of farm sites in which species occurred) in the second snap bean crop included *Lamium amplexicaule*, *Mollugo verticillata*, *Solanum ptychanthum*, *Stellaria media,* and *Ipomoea hederacea*. The species with the highest density included *Mollugo verticillate* in the first snap bean crop and *Stellaria media* in the second snap bean crop. Future work will examine regional variation in snap bean weed communities and whether there are associations with agronomic practices.

POTENTIAL HERBICIDES TO CONTROL PROBLEM WEEDS IN SNAP BEANS. B. Scott*,
University of Delaware, Georgetown, DE

ABSTRACT

Snap beans (*Phaseolus vulgaris*) are grown for processing and fresh market in Pennsylvania, Maryland and Delaware. Snap beans can compete fairly well with weed pests but it is important to keep fields weed free until beans reach canopy closure. Few herbicides are labeled for weed control and additional products are needed to improve overall control. This study was initiated to evaluate potential PRE herbicides for crop injury and weed control in snap beans.

Trials were arranged as a randomized complete block design with three replications. Fields were conventionally tilled and 'Caprice' snap beans were planted on 30 inch rows at 6 seeds per ft-row. Snap beans were planted on June 20 on loamy sand in DE, June 6 on silt loam in PA, May 30 on silt loam in MD. PRE herbicides were applied one day after planting and POST herbicides were applied at the second trifoliate at all locations. All treatments included PRE *S*-metolachlor at 1.19 lb ai/A in DE or 1.59 lb ai/A in PA and MD. Treatments that included halosulfuron were applied at 0.023 lb ai/A in DE or 0.031 lb ai/A in PA and MD. PRE treatments included *S*-metolachlor + each of the following: flumioxazin at 0.032 and 0.064 lb ai/A, sulfentrazone at 0.094 and 0.188 lb ai/A, lactofen at 0.188 lb ai/A, oxyfluorfen at 0.25 lb ai/A, fomesafen at 0.313 lb ai/A, and halosulfuron. PRE followed by (f/b) POST treatments included *S*-metolachlor PRE plus the following: flumioxazin at 0.064 lb ai/A f/b bentazon at 0.75 lb ai/A + fomesafen at 0.188 lb ai/A, sulfentrazone at 0.188 lb ai/A f/b bentazon + fomesafen, lactofen at 0.188 lb ai/A f/b bentazon + fomesafen, halosulfuron f/b bentazon + fomesafen, and halosulfuron f/b bentazon + imazamox at 0.31 lb ai/A. An untreated check was also included. In DE, a POST application of bentazon was added to PRE treatments to control patches of common cocklebur. All POST treatments included non-ionic surfactant. Snap bean injury and weed control were evaluated visually at 4, 6, and 8 weeks after planting (WAP) based on a 0 to 100 scale. Snap beans were harvested and weight was recorded. Data were analyzed and means were separated using Fisher's Protected LSD ($\alpha = 0.05$).

At 6 WAP minimal crop injury was observed with halosulfuron alone, fomesafen alone, halosulfuron f/b fomesafen and halosulfuron f/b imazamox in both DE and PA. In MD, 0-7% mid-season injury was observed with all treatments except oxyfluorfen. The DE best yielding treatments were: halosulfuron f/b imazamox, halosulfuron f/b fomesafen, the low rate of sulfentrazone, respectively. The PA best yielding treatments were: halosulfuron alone, fomesafen alone, halosulfuron f/b fomesafen and halosulfuron f/b imazamox, respectively. In DE the high rate of flumioxazin, sulfentrazone and treatments with fomesafen provided good Palmer amaranth control (≥89%). No treatment provided adequate season long morningglory control. In PA excellent weed control was observed with smooth pigweed, ladysthumb, and giant foxtail. Common lambsquarters control was slightly reduced with lactofen and fomesafen alone although control was still 90% or greater. Good to excellent common ragweed control was observed with all treatments except sulfentrazone alone and oxyfluorfen. Across three states and two soil types two treatments resulted in minimal injury, higher yields and effective weed control for most species: 1) halosulfuron + *S*-metolachlor f/b fomesafen + bentazon and 2) halosulfuron + *S*-metolachlor f/b imazamox + bentazon.

BICYCLOPYRONE- CROP SAFETY AND WEED CONTROL IN MINOR CROPS. L. Smith*, T. Trower, T. Beckett, and G. Vail, Syngenta, King Ferry, NY

ABSTRACT

Bicyclopyrone is a HPPD-inhibiting (Group 27) herbicide developed by Syngenta and registered for broadleaf and grass weed control in field, sweet and popcorn. University and Syngenta studies have evaluated the crop safety and weed efficacy of preemergence and postemergence bicyclopyrone applications in various minor crops. Preemergence applications of 37.5 to 50 g ai/ha have exhibited good to excellent crop tolerance and acceptable broadleaf weed control in many of the crops tested. Syngenta is pursuing use labels on timothy grown for seed, watermelon, horseradish and strawberry. Preemergence applications for onion grown in muck soils is also under evaluation.

PREEMERGENCE GRANULAR FORMULATION EFFICACY COMPARISONS. A.R. Shiffer*, J. Neal, and C. Harlow, North Carolina State University, Raleigh, NC

ABSTRACT

Early research conducted showed the efficacy of Marengo G was reduced when the formulation was changed from the 'ECO' granule, a standard clay granule, to the 'Verge' granule. Other products have also changed their standard formulation from a clay to 'Verge' granule. This experiment was conducted to deduce if a change in efficacy occurred when this change in formulation was made. Experimental plots of four-liter plastic pots were treated with pre-emergence granular herbicides then seeded with common nursery weeds. At 4 weeks after treatment, weed seedlings with at least 2 true leaves were counted, and approximately every 2 weeks thereafter control was visually evaluated as % control. Evaluations on granular pre-emergent efficacy concerning formulation changes from a standard clay granule to 'Verge' granule yielded results indicating that the new formulation does not lower or substantially change the efficacy of the product, except in the case of Marengo G. Marengo G 'Verge' consistently underperformed against Marengo G 'ECO' in controlling common nursery weed species.

BENTAZON INJURY TO SEDUM. C. Harlow* and J. Neal, North Carolina State University, Raleigh, NC

ABSTRACT

Sedum, or stonecrop, is an easy to grow group of succulents which continues to increase in popularity in landscape use. Few herbicides are labeled for weed control in sedum production, with no postemergence herbicide currently labeled for over-the-top broadleaf weed management. The objective of this study was to evaluate the safety of bentazon (Basagran T&O 4L) herbicide applied over-the-top to five species of container-grown sedum. The experiment was conducted as a RCBD with 4 replicates. Basagran T&O was applied over-the-top of Kamchatka stonecrop, two-row stonecrop, tasteless stonecrop, Jenny's stonecrop, and white stonecrop. Rooted liners of all species were potted into 3-L pots on June 21, 2019 and placed on a nursery pad to establish. Herbicides were applied September 3, 2019 with a second application 15 days later, according to label directions. Basagran at 32-, 64-, or 128 oz/A was applied with a CO_2-pressurized bottle sprayer at 30 GPA using flat-fan nozzles. Nontreated check plants were included for comparison. Plant injury was visually evaluated on a 0 to 10 scale where 0 = no visible injury and 10 = dead plants. Additionally, fresh weights of plants were measured at the termination of the experiment. Severe injury to all *Sedum* species was observed at all doses, beginning 2 days after the initial herbicide treatment. Injury was primarily in the form of leaf necrosis, with some discoloration and stunting of plants, as well. Injury was greater at higher doses and became worse after the second application of treatments. Mortality was observed in many of the plants. For some species, plants that survived eventually began to recover slightly; however, overall injury was severe. Based on these results, Basagran T&O should not be used over-the-top on these stonecrop species.

GEMINI G HERBICIDE SAFETY TO *SEDUM ALBUM* AND *SEDUM RUPESTRE*. J.S. Aulakh*, The Connecticut Agricultural Experiment Station, Windsor, CT

ABSTRACT

Gemini G (isoxaben 0.25% + prodiamine 0.40%) is a granular herbicide for preemergence control of several broadleaf weeds and grasses in container, field grown, and landscape ornamentals. Overhead irrigated container trials were conducted in 2018 and 2019 at the Valley Laboratory of the Connecticut Agricultural Experiment Station at Windsor, CT. Tolerance of Jenny's stonecrop (*Sedum rupestre*) variety 'Blue Spruce' and white sedum (*Sedum album*) variety 'Red Ice' to two sequential applications (6 wk apart) of Gemini G herbicide was evaluated. Experiments were conducted in a completely randomized design with 12 plants per treatment. Sedum plugs (15 cm) were transplanted into 1-gallon pots on June 27, 2018 and June 10, 2019. Gemini G was applied over-the-top with a shaker bottle at 0, 200, 400, and 800 lb/ac on July 4, 2018 and June 17, 2019 and again on August 15, 2018 and July 29, 2019. In 2018, no injury was observed at any stage of growth with Gemini rates up to 800 lb/a. However, in 2019, significant stunting injury (0-10 scale) occurred in both species following the second application. Maximum stunting injury occurred 2 wk after the second application and was rated 3.1 and 2.2 with 800 lb/a and 400 lb/a, respectively. Based on these results it appears Gemini G rates up to 400 lb/a can safely be applied for preemergence weed control in both sedum species if minor stunting injury can be tolerated.

LEGACY EFFECTS OF CROP DIVERSITY ON SOIL-MEDIATED WEED-CROP COMPETITION.
U.D. Menalled*, M.R. Ryan, A. DiTommaso, and R. Smith, Cornell University, Ithaca, NY

ABSTRACT

Researching the interaction between diversity and plant competition can promote ecological weed management. Ecological theory postulates that increased plant diversity facilitates the exploitation of more ecological niches. Subsequent agronomic research showed that greater cropping diversity reduced weed-crop competition. We tested whether weed-crop competition was affected by cropping diversity legacy. Four levels of crop diversity were grown in annual and perennial systems from spring 2016 to spring 2019 in Aurora, NY and Alburgh, VT. A corn monoculture was seeded at both sites immediately after moldboard plowing the diversity treatments in spring 2019. During the ensuing growing season, each of the diversity conditioning treatments was partitioned into weed-free, low-weed, high-weed, and ambient weed treatments. Sunflower was seeded as a surrogate weed to establish the low and high-weed treatments at 34,790 and 69,580 seeds/ha., respectively. Weeds from the soil seedbank grew in the low and high-weed treatments as well as an ambient weed treatment. Weed communities in the ambient treatment were sampled by species and described with a Bray-Curtis dissimilarity matrix. Subsequent visualization of the Bray-Curtis matrix through NMDS and analysis with a PERMANOVA showed that weed community structure differed between sites and was affected by the system (annual vs. perennial) legacy of the soil. To elucidate weed-crop competition in each of the soil legacies, corn and weed biomass was modeled through an inverse rectangular hyperbola. Analysis of the competition curves suggested that, at both sites, neither conditioning system nor diversity had legacy effects on weed-crop competition. As such, this work infers that crop diversity effects on weed-crop competition may have weak legacies that can be reduced by primary tillage events.

A CHANGE IN WEED-CROP COMPETITION THROUGH MICROBIALLY MEDIATED NITROGEN IMMOBILIZATION. M.A. Gannett*, J. Kao-Kniffin, and A. DiTommaso, Cornell University, Ithaca, NY

ABSTRACT

There is a need for non-chemical based weed management tools since broad use of herbicides has created a strong selection pressure on weed communities and a subsequent increase in herbicide resistant weeds. A potential method of weed control is through managing soil nitrogen availability. It is hypothesized that soil carbon additions stimulate soil microbial growth, which increase nitrogen immobilization. Many weeds that grow well in high nitrogen environments will have reduced growth when more nitrogen is immobilized. In two experiments we added carbon amendments to the soil and predicted that it would affect weed biomass. Microbial biomass was measured from pots buried in a field, either unamended or amended with sawdust and sugar as the carbon source. In carbon amended pots, microbial biomass was significantly greater. Additionally, there was more nitrate leached through the soil column into resin bags in the unamended pots, providing supporting evidence that more nitrogen was immobilized in carbon amended pots. Weed biomass for three of the eight weed species grown in the pots had significantly less biomass in carbon amended pots after 11 weeks. These species are known to grow well in high nitrogen environments. Similar results were found in a soybean field experiment with three treatments: unamended plots, low carbon level sawdust amended plots, and high carbon level straw amended plots. Straw amended plots had significantly less aboveground weed biomass. There was no effect of amendment on soybean aboveground biomass, which is a nitrogen fixing legume. Together, these experiments support the hypothesis that nitrogen immobilization through carbon additions could be developed into an alternative weed management tool.

MICROBIAL CONTRIBUTIONS TO WEED SUPPRESSION IN CONVENTIONAL AND ORGANIC FARM SOILS. J. Kao-Kniffin, A. DiTommaso, and L. CHENG*, CORNELL UNIVERSITY, Ithaca, NY

ABSTRACT

The recent advances in sequencing technologies could provide insights into the complex interactions between weed species and soil microbiota that influence weed growth. Specifically, we collected soil samples from common ragweed (*Ambrosia artemisiifolia*) in 24 locations with different cropping systems in New York State, and examined plant-soil feedback effects in a greenhouse experiment. Microbiomes from the 24 farm soils were added to replicated pots containing ragweed seedlings. Mature plants were removed and the soils were re-planted with new seedlings to simulate a plant-soil feedback cycle. A machine-learning algorithm was constructed to create a supervised principal components analysis (PCA) of bacterial 16S rRNA gene sequences derived from Pearson correlations linking microbial taxa with weed growth. Distinct microbial fingerprints emerged that separated conventional and organic cropping systems by weed suppression level. A large proportion of the most highly suppressive microbiomes were derived from conventional farms, whereas the microbiomes resulting in positive growth, neutral, or weak suppression of ragweed originated largely from organic farms. The sequencing data revealed that levels of negative plant-soil feedback were influenced by farm management. Network analysis showed completely different bacterial interaction networks between organic and conventional farms. Our results suggest that the soil microbiota associated with surviving populations of ragweed inhibits the growth of the successive cycle of ragweed plants in conventional farms. Further investigations of these highly suppressive microbiomes using laboratory cultivation techniques and activity-based metagenomics could reveal specific biological agents and natural products that may be suitable for weed management.

ECOTYPIC STRESS RESPONSE DIFFERENCES IN AN INVASIVE GRASS ARE DRIVEN BY HOME CLIMATE AND HABITAT. V.T. Lakoba* and J.N. Barney, Virginia Tech, Blacksburg, VA

ABSTRACT

Species invasion continues to accelerate across the planet, largely facilitated by increasing global interconnectivity. In building a better understanding of invasion and our response to it, we must consider humans' role in shaping invasions through land management. How do climate and land use impact local adaptation in invasive species? To address this question, we tested climate and ecotype (agricultural vs. non-agricultural habitat) as drivers of rapid divergent evolution in the invader Johnsongrass (*Sorghum halepense* (L.) Pers.). We hypothesized ecotypic differences based on a body of previous work which found divergence in plasticity and response to competition, as well as an understanding of agricultural populations' release from the selection pressures of drought and nutrient stress. We subjected five agricultural populations and five non-agricultural populations sampled from across the United States to nutrient- and/or drought-stress treatments during the early establishment life stage. We found that the non-agricultural populations produced more biomass, suffered less drought impact, and showed adaptation across a spectrum of soil fertility. Environmental factors at population origins also impacted growth response, both individually and in tandem. We present evidence of an emergent non-agricultural ecotype which is more vigorous and stress tolerant than its agricultural counterpart. This ecotype's emergence is largely due to successful control of the invader in agricultural systems in the latter twentieth century. We foresee further studies into ways that invasion management can contribute to evolutionary trajectories.

CRITICAL PERIOD OF ITALIAN RYEGRASS CONTROL IN NO-TILL WINTER WHEAT SYSTEMS. D.J. Contreras* and W.J. Everman, North Carolina State University, Raleigh, NC

ABSTRACT

Winter wheat (*Triticum aestivum*) is often used in double cropping systems, as it can aid reducing soil erosion, weed suppression, and provides growers with an additional source of income. Italian ryegrass (*Lolium multiflorum*) is one of the most troublesome weeds for wheat production throughout Southern US. The inability to control this weed can result in reduced yields, reduced quality, or both. Cultural practices, such as the Critical Period of Weed Control (CPWC) can help reduce yield losses. The CPWC is a period in a crop´s growth cycle during which weeds must be controlled to prevent yield losses. The objective of this study was to determine winter wheat's CPWC for Italian ryegrass that results on or less than 5% total yield loss. On this two-year study, plots were kept weedy or weed free throughout two weekly intervals (Year 1) and three-week intervals (Year 2) after crop emergence. Two controls consisting of weedy and weed free all-season plots were used for a treatment comparison based on a regression analysis where time of weed removal was related to crop yield. The CPWC to reduce yield losses to 5% or less on Year 1 was found to start around the 7th WAE (6.89, $p \leq 0.05$) and end at 19th WAE (19.36, $p \leq 0.05$). On Year 2, the CPWC started on the 11th WAE (11.00, $p \leq 0.05$) and finished at 23rd WAE (23.16, $p \leq 0.05$). Factors such as earlier planting dates, warmer weather and precipitation changes can give way to such variations in timing of the CPWC, but its duration was maintained.

UTILIZING MULTISPECTRAL AND HYPERSPECTRAL SENSORS TO DISCRIMINATE MULTIPLE WEED SPECIES. M. Granadino*, W.J. Everman, J.T. Sanders, E.A. Jones, and M.A. Fajardo, North Carolina State University, RALEIGH, NC

ABSTRACT

For unmanned aerial vehicle (UAV) remote sensing to be employed as a viable and widespread tool for weed management, the accurate detection of distinct weed species must be possible using analytical procedures on the resultant imagery. In 2019, two field studies were performed to identify any weed height thresholds on the accurate species detection and species by height classification of three common broadleaf weed species in Rocky Mount, North Carolina: Palmer amaranth (*Amaranthus palmeri*), common ragweed (*Ambrosia artemisiifolia*) and sicklepod (*Senna obtusifolia*). Pots of the three species at heights of 2, 6 and 12 cm were randomly arranged in a grid and 5-band multispectral imagery was collected at 15 m above the ground. Image analysis was used to identify the spectral reflectance behavior of the weed species and height combinations and to evaluate the accuracy of species based supervised classifications. Supervised classification was able to discriminate between the three weed species with between 20-100% accuracy depending on height and species. Palmer amaranth classification accuracy was improved at larger weed heights and detection accuracy for the species was at least 68%. Increased height of sicklepod and common ragweed plants did not reliably confer improved accuracy, but the species were correctly identified with at least 68% and 40% accuracy, respectively. Spectral separation was found to be dependent on the species, band, and height being observed. Reflectance of the species in bands 1, 2, 4, and 5 increased with height while band 3 reflectance declined as weed height increased. Finally, spectral separation was found to exist between the species, demonstrating the potential for using spectral reflectance measurements to discriminate between them.

ASSESSING GLUFOSINATE USE AND EVALUATING FARMER USE EFFICACY IN NORTH CAROLINA SOYBEANS. E.A. Jones*, W.J. Everman, R.G. Leon, and C. Cahoon, North Carolina State University, Raleigh, NC

ABSTRACT

Glufosinate resistance has yet to evolve in any broadleaf weed globally. However, since many broadleaf weeds, such as *Amaranthus palmeri* (Palmer amaranth) exhibit multiple herbicide resistance, glufosinate may be applied more extensively to control herbicide-resistant weeds in crop fields. If this is indeed what farmers will do for weed control, the evolution of glufosinate resistance is inevitable. Determination of how farmers use glufosinate was assessed by administering a survey at the row crop extension meetings in the winter of 2019. The results of the survey indicated that farmers are using glufosinate for resistance management (48%), while some farmers were solely relying on glufosinate for weed control (17%). The surveyed farmers also responded that glufosinate was primarily applied at EPOST (25%), while POST (10%) and EPOST+POST (15%) was applied frequently as well.

The descriptive statistics were used to construct an experiment to determine the control and fecundity of glufosinate-treated Palmer amaranth in soybeans. The herbicide programs included applications of glufosinate at the following timings: EPOST, POST, LPOST; EPOST followed by (fb) POST; EPOST fb LPOST; POST fb LPOST; EPOST fb POST fb LPOST; nontreated. EPOST (98%) was the most efficacious single application followed by POST (73%) then LPOST (53%). Herbicide programs with two or three glufosinate applications controlled at least 90% of weeds.

Palmer amaranth plants survived and remained fecund after glufosinate applications of POST, LPOST, EPOST fb LPOST, and POST fb LPOST. The nontreated Palmer amaranth plants produced 2000 seeds per plants. Plants surviving the POST and LPOST glufosinate applications produced the same amount of seeds as the nontreated plants. Plants surviving the EPOST fb LPOST and POST fb LPOST programs produced seeds per 700 and 173 seeds per plant, respectively.

Overall, weed control was the most efficacious at EPOST and no seeds were produced. Visual evaluations of weed control was high with the sequential applications of glufosinate, but some plants that were not controlled produced seeds. Farmers should remain applying glufosinate at an EPOST timing. However, the results of the experiment provide clear evidence that weeds can survive while remaining fecund with some application timings of glufosinate. Thus, farmers must realize that the evolution of glufosinate resistance will be accelerated if applications are untimely and no other control tactics are utilized.

ANNUAL BLUEGRASS WEEVIL (*LISTRONOTUS MACULICOLLIS*), OVERSEEDING, AND PACLOBUTRAZOL FOR ANNUAL BLUEGRASS (*POA ANNUA*) CONTROL IN GOLF FAIRWAYS. K.H. Diehl*, M.T. Elmore, J.A. Murphy, A.M. Koppenhöfer, and O.S. Kostromytska, Rutgers University, New Brunswick, NJ

ABSTRACT

Annual bluegrass (*Poa annua* L.) is one of the most problematic weeds of cool-season turfgrass. The annual bluegrass weevil (*Listronotus maculicollis*; ABW) is a turfgrass pest that showcases ovipositional preference for annual bluegrass over creeping bentgrass (*Agrostis stolonifera* L.). Turfgrass managers typically use insecticides to control ABW preventatively and avoid turfgrass damage. We hypothesized that withholding ABW insecticides until a damage threshold was met would reduce annual bluegrass cover in creeping bentgrass fairways. The effect of ABW damage was tested alone and in combination with monthly applications of paclobutrazol and creeping bentgrass overseeding. Research was conducted from 2017 to 2018 on a simulated creeping bentgrass fairway infested with annual bluegrass at the Rutgers Horticulture Farm No. 2 in North Brunswick, NJ. Three insecticide programs were evaluated in combination with paclobutrazol and overseeding in a 3 x 2 x 2 complete factorial. Treatments were replicated four times and arranged in a split-split plot design with overseeding as a whole plot factor, insecticide program as the sub-plot factor, and paclobutrazol as the sub-sub plot factor. Paclobutrazol (280 g ha^{-1}) was applied monthly from May through October. The first insecticide program (preventative) was an industry standard designed to control early ABW larval stages and prevent turfgrass injury. In the second program (threshold), no insecticides were applied until visual ratings determined that ABW damage resulted in unacceptable turfgrass quality. Once this damage threshold was met, an insecticide was applied to prevent future turfgrass injury. No insecticides were applied to the third program (no-insecticide) for ABW control. Whole plots were overseeded with '007' creeping bentgrass (49 kg ha^{-1}) on 12 June 2017 and 2018, approximately one week after the damage threshold was met. To evaluate the efficacy of the insecticide programs, cores (6 cm in diameter) were taken from each plot and larvae were extracted in salt solution. Extracted larvae were counted and used to determine the average larval density per plot. Turfgrass quality, annual bluegrass quality and creeping bentgrass quality were evaluated monthly on a 1 (poor) to 9 (excellent) scale where 6 is considered acceptable. Lightbox photos were taken monthly and subjected to digital analysis (Turf Analyzer software). Percent annual bluegrass cover was evaluated visually each month and grid intersect counts were taken in November. Data were subjected to ANOVA in SAS (v9.4) as a split-split plot design using the MIXED procedure (α=0.05) and Fisher's Protected LSD (α=0.05) was used to separate means.

The main effect of paclobutrazol was significant for annual bluegrass cover from May 2017 through April 2019. The main effect of insecticide program was significant in July 2017 and from September 2018 until the experiment was concluded in April 2019. Interactions between main effects were observed in July and August in 2018, and on the final rating sat in April 2019. At the conclusion of this two-year experiment in April 2019, paclobutrazol treated plots had 0% annual bluegrass cover. In the absence of paclobutrazol, annual bluegrass cover was lower in the no-insecticide program (23%), than the threshold and preventative programs (37-44%) at the conclusion of the experiment. Overseeding did not affect annual bluegrass cover either year.

THE IMPORTANCE OF DAY LENGTH IN MODELING *RAPHANUS RAPHANISTRUM* PHENOLOGY. T.A. Reinhardt Piskackova*, S. Reberg Horton, R.J. Richardson, K.M. Jennings, and R.G. Leon, North Carolina State University, Raleigh, NC

ABSTRACT

Raphanus raphanistrum is a common weed of winter crops around the world; however, growth through the winter months can present difficulties for predictive modeling. In this study, three cohorts of *R. raphanistrum* were tracked to create predictive phenology models. When accounting for day of emergence, all three cohorts were able to be modeled together. Different environmental parameters were important to incorporate for certain phenological stages. Thermal time might be adequate to predict the vegetative stages of *R. raphanistrum*, but day length was important to predict reproductive stages. This information is important to understand *R. raphanistrum* biology and to time diverse management tactics for multiple cohorts that may emerge within winter crops.

COMPARING THE ROLE OF SOIL WATER POTENTIAL AND HYDRAULIC CONDUCTIVITY ON WEED SEED GERMINATION. M.E. Camacho*, T.W. Gannon, A. Amoozegar, J.L. Heitman, and R.G. Leon, NCSU, Raleigh, NC

ABSTRACT

Seed germination is probably the one of the most important processes in weed biology and ecology, due it determines the timing, the number of individuals and the level of weed competition with crops. Several approaches have been performed for modelling this process aiming to understand the dynamics of seeds under different environmental factors, and provide useful information for proper management. Seed hydrothermal time approach is considered effective while describe the seed physiologic response to changes environmental conditions, including temperature and water potential. However, methods for this approach are performed in laboratory conditions, using polyethylene glycol (PEG) as germination media due its versatility to set different water potential values.

Present work was carried out to compare seed germination of four plant species in PEG and four soils with contrasting textures under six water potentials, aiming to evaluate the performance of this polymer representing the edaphic conditions for further field management.

As a result, we obtained that total seed germination showed a trend to decrease with a drop in the water potential, with differences between soil and PEG under the same water potential values. Seeds of *Triticum aestivum* presented values higher than 93% irrespective of water potential, while germination rapidly decreased as the water potential became more negative in the four soils. Due this inconsistency associated with soil water potential and total germination, the role soil hydraulic conductivity (K_h) was evaluated, and presented a better performance in explaining the germination trends. In general, total germination reached the maximum values within the highest K_h. As the hydraulic conductivity decreases (mainly due soil K decreases as soil water potential decreases) until achieve a critical value, the germination decreased sharply. As well, we found the rate to reach maximum germination (GR_{max}) was sensitive to changes in K_h, getting unique responses depending on the seed species and the soil texture.

Our finding suggest that PEG does not represent faithfully most of soil conditions while evaluating under the same water potential values, probably due the difference soil and PEG in seed media contact area and the components of soil total water potential. Soil hydraulic conductivity became as a useful variable to predict changes in both seed total germination and germination rate. Special attention should be pay to results obtained using PEG as germination media before further use in field practices and research.

EVALUATING HERBICIDES FOR POSTEMERGENCE CONTROL OF COMMON WEEDS IN CONTAINER NURSERY CROPS.

H. Lin*, J. Neal, and C. Harlow, North Carolina State University, Raleigh, NC

ABSTRACT

Studies were conducted in winter 2018 and summer 2019 in North Carolina to evaluate herbicides for postemergence control of common annual weeds in container nursery production. Plastic pots were filled with hammer-milled pine bark substrate amended with 8 lb/yd^3 lime and 10 lb/yd^3 of Harrell's 18-4-8 slow-release fertilizer. The winter study was conducted at NCSU's Horticulture Field Lab (HFL), Raleigh, NC. Summer studies were conducted at HFL and NCSU's Horticultural Crops Research Station in Castle Hayne (CH), NC. Weed species for the winter study were common groundsel (*Senecio vulgaris)*, common chickweed (*Stellaria media)*, flexuous bittercress (*Cardamine flexuosa)*, and yellow woodsorrel (*Oxalis stricta)*. For the summer studies, chamberbitter (*Phyllanthus urinaria*), eclipta (*Eclipta prostrata*), livid amaranth (*Amaranthus lividus*), rice flatsedge (*Cyperus iria*), spotted spurge (*Euphorbia maculata*), and yellow woodsorrel (*Oxalis stricta*) were included. All weeds were surface seeded twice, two weeks apart, to evaluate the effects of growth stage on herbicide efficacy. Treatments included diquat at 0.56 kg ai/ha, flumioxazin at 0.42 kg ai/ha, flumioxazin + pyroxasulfone at 0.532 kg ai/ha, halosulfuron-methyl at 0.0526 kg ai/ha, indaziflam at 0.049 or 0.0653 kg ai/ha, isoxaben at 1.12 kg ai/ha, sodium salt of bentazon at 0.84 or 1.12 kg ai/ha (only 1.12 kg ai/ha was included in the summer), sulfentrazone at 0.28 or 0.56 kg ai/ha (only 0.28 kg ai/ha was included in the summer), and a nontreated control. Additionally, cloransulam-methyl at 0.0177 kg ai/ha, glufosinate at 1.12 kg ai/ha, oxyfluorfen at 0.56 kg ai/ha, saflufenacil at 0.025 kg ai/ha, and topramezone at 0.196 kg ai/ha were included in the summer studies. Percent weed control was visually evaluated weekly for 4 and 6 weeks after application for summer and winter studies, respectively, on a 0 to10 scale where 0 = no control (no difference from the nontreated plants) and 10 = complete (100%) control. Fresh weights were measured at 4 or 7 weeks after treatments for summer and winter studies, respectively. In the winter study, diquat, flumioxazin, and flumioxazin + pyroxasulfone provided ≥ 80% control of common groundsel, common chickweed, flexuous bittercress, and oxalis for both growth stages. In summer studies, flumioxazin provided ≥ 80% control of chamberbitter, eclipta, livid amaranth, rice flatsedge, and yellow woodsorrel. Glufosinate provided ≥ 80% control of chamberbitter, eclipta, and spotted spurge.

EVALUATING CROP RESPONSE OF VARIOUS VEGETABLE SPECIES TO SIMULATED DICAMBA DRIFT. M. Wasacz*, M.J. VanGessel, D.J. Mayonado, and T.E. Besancon, Rutgers University, New Brunswick, NJ

ABSTRACT

The spread of herbicide resistant Palmer amaranth (*Amaranthus palmeri* S. Watson) and horseweed (*Erigeron canadensis* L.) has recently contributed to increase the acreage of dicamba-tolerant soybean in the mid-Atlantic region. Dicamba is a synthetic auxin herbicide that is prone to volatilization and drift, which may put neighboring dicamba-sensitive crops at risk. Our research seeks to determine the relative sensitivity of some vegetable crops species frequently grown in the mid-Atlantic region to sublethal rates of dicamba in comparison to sensitive soybean. Vegetable crops in this experiment included tomato (*Solanum lycopersicum*), bell pepper (*Capsicum annuum*), eggplant (*Solanum melongena*), sweet basil (*Ocimum basilicum*), summer squash (*Cucurbita pepo*), watermelon (*Citrullus lanatus*), cucumber (*Cucumis sativus*), pumpkin (*Cucurbita pepo*), garden bean (*Phaseolus vulgaris*), lima bean (*Phaseolus lunatus*), and kale (*Brassica oleracea*). In order to simulate exposure to dicamba drift, all species were sprayed with following fractions of the dicamba recommended label rate (560 g a.e. ha^{-1}): 1/250, 1/500, 1/1000, 1/2000, 1/5000, 1/10000, and an untreated control. Height, number of leaves, and visual injury ratings, including chlorosis, necrosis, stunting, epinasty, and leaf deformation, were taken 1, 2 ,3 and 4 weeks after treatment (WAT). Our preliminary results suggest that tomato, eggplant, lima bean, and garden bean were the most sensitive to dicamba with at least 20% leaf cupping 3 WAT at rates above 1/1000 of the labeled rate, and still significant injury at 1/2000 or 1/5000 of the labeled rate. Lack of tolerance to dicamba of these crops was similar to what has been observed with sensitive soybean. Bell pepper, cucumber, and watermelon were slightly more tolerant with over 20% leaf cupping at 1/250 of the labeled rate, but minimal to no injury at lower rates. Basil, kale, and summer squash were the least sensitive crop with less than 10% leaf cupping 3 WAT when exposed to the highest dicamba rate (1/250), and virtually no injury at rates equal or less than 1/5000 of the recommended rate. Future studies will seek to more accurately quantify the degree of leaf cupping in response to dicamba sublethal rates, as well as examine the translation of leaf injury into potential yield reduction.

EFFECTS OF DIFFERENT PLANTING DATES AND SUB-LETHAL RATES OF DICAMBA ON SOYBEAN. M.A. Fajardo* and W.J. Everman, North Carolina State University, Raleigh, NC

ABSTRACT

Production of soybean (*Glycine max*) in USA is one of the main agricultural interest, being the country leader in production and falling to second place in terms of exporting this grain. The use of soybeans can mainly be separated into the largest source for animal protein feed and processing for vegetable oil, being soybeans the second source for this and in the country representing the 90 percent of oilseed production. The use of new and better technologies has changed the rules of planting dates and the correlation it has in terms of the crop yield by the end of the season. Different varieties and maturity groups planted early may become an improved technique farmers may use as advantage with better results this past years. Being a root or foliage absorbed systemic herbicide Dicamba may be a potential injury factor in soybeans, affecting the growing points of the plant and reducing by this the plant full capacity of developing, affecting yield according to the rate of exposure. The research had for objective to assess the effect in terms of injury and yield in two different maturity groups, each with two different soybean varieties, when exposed at vegetative (V4) or reproductive stages (R2) and with two different planting dates (May or June). The location of the research was the Upper Coastal Plain Research Station at Rocky Mount, NC. Rates of Dicamba applied were 0.0, 1.08, 4.35, and 17.40 g ae ha-1 to soybeans. Evaluations to the treatments were conducted at 14, 21, 28, 35 and 42 days after treatment (DAT). Results provide evidence that early or late planting dates do not have significant effects on different evaluation timings in terms of injury in soybeans.

INTERCROPPING PERENNIAL GRAINS AND PEAS: CHALLENGES AND OPPORTUNITIES. E.P. Law*, C.J. Pelzer, S. Wayman, M.R. Ryan, and A. DiTommaso, Cornell University, Ithaca, NY

ABSTRACT

Diversified cropping systems that include polycultures of perennial and annual crops and provide enhanced ecosystem services in addition to producing food, forage, fiber, and fuel have the potential to greatly contribute to solutions to the intractable problem of feeding a growing human population without further degrading the natural capital that drives agriculture. Here we report on a two-year field experiment investigating the viability of intercropping two perennial small grain crops with food-grade field pea (*Pisum sativum*), an annual legume. The two perennial grains, 'Kernza®' intermediate wheatgrass (*Thinopyrum intermedium*; IWG) and 'ACE-1' perennial cereal rye (*Secale cereale* x *S. strictum*; PR) are currently being developed for dual-use grain and forage production in New York. The experiment was planted September 2017 using a spatially balanced complete block design with IWG, PR, and field pea monocultures, and IWG/pea and PR/pea polycultures as main plot treatments. A winter-killed nurse crop of oats (*Avena fatua*) was also applied as a split-plot treatment to explore its potential for weed suppression during perennial grain establishment. Crop and weed biomass were sampled from two 0.5 m^2 quadrats in each split-plot in 2018 and 2019. Pea growth was severely inhibited in polyculture with either perennial grain, and land equivalent ratios and harvest incompatibility indicate that intercropping these species is likely not a viable strategy. PR grain yields averaged between 2300 and 3000 kg/ha in 2018, with higher yields produced in monoculture, but dropped to 200 to 400 kg/ha in 2019 likely due to high weed pressure after the first harvest. IWG grain yields were consistently between 300 and 500 kg/ha in both years. The two perennial crops exhibited opposite trends in total biomass production, with PR biomass declining from averages of 7000 and 9600 kg/ha in 2018 to 1700 and 2900 kg/ha in 2019 in polyculture and monoculture, respectively, while IWG biomass increased from 3000 and 3700 kg/ha in 2018 to 6600 and 7400 kg/ha in 2019. The nurse crop treatment had no impact on any variables measured in PR plots, but was surprisingly associated with reduced weed biomass in IWG/pea polyculture plots in 2019. These results provide important insights into the life histories of these two crops that have implications for both production management and further breeding and agronomic development efforts. We suggest that research efforts in PR focus on improving persistence, competitive ability, and post-harvest weed management, while research in IWG might focus on weed management during establishment and efforts to increase yield through further breeding and improvements in agronomic factors such as fertilization and seeding rate.

CANOPY RESPONSES TO HOMOGENEOUS AND HETEROGENEOUS PLANTING ARRANGEMENTS IN CORN. S.A. Ramsey*, W.J. Everman, D.L. Jordan, A.M. Locke, and R.G. Leon, North Carolina State University, Raleigh, NC

ABSTRACT

There is a need for cultural and mechanical practices in row crop agriculture to control herbicide-resistant (HR) weed populations. An important limitation for developing effective integrated weed management, especially with the HR populations, is the lack of strategies to prevent or at least reduce the production of new seed by weeds surviving herbicide application or cultivation. The objective of this study is to identify planting arrangements in corn that increase weed suppression and protect crop yield under weedy conditions and increase relative crop yield under weed free conditions. Through the use of the solar corridor concept, we have created four different planting arrangements in order to maximize light interception to lower layers within the crop canopy and increase crop yield. We have found that in a uniform, high density planting of corn, there is a negative effect on both plant growth and quality of ears. A uniform, single density planting arrangement had the highest amount of dry weight accumulation at the end of the season. A patterned planting arrangement had greater leaf area index than a single planting density arrangement in the beginning of the season. These results show that corn has been intensively bred in order to avoid etiolation and to reduce the plant's plasticity at varying densities. In the future, using crops with more plasticity could increase the significance of the planting arrangements.

NON-CHEMICAL WEED CONTROL IN AMENITY GRASSLANDS VIA FRAISE MOWING. C.R. Sitko* and F.S. Rossi, Cornell University, Ithaca, NY

ABSTRACT

Responding to public pressure to reduce synthetic chemical use and combating the rise of herbicide resistance in amenity grasslands are challenging due to the lack of alternative options. Additionally, the changing climate is less conducive to traditional grassland species success. The ability to rapidly renovate grassland surfaces could provide solutions to both challenges by reducing weed seed pressure and establishing genetically improved, well adapted varieties. Fraise mowing is an aggressive cultivation practice designed to harvest verdure, organic matter, and soil to a depth of 5 cm while allowing turfgrass to reestablish from unharvested rhizomes or provide an advantageous seedbed for establishing via seed or sod. Given the weed seed bank accumulates near the soil surface in no-till systems, we hypothesized fraise mowing could be an effective means of weed seed bank harvest and long-term, non-chemical weed control. Field experiments were conducted from 2017 – 2019 at the Vineyard Golf Club (Edgartown, MA) in naturalized fine fescue rough heavily infested with smooth crabgrass (*Digitaria ischaemum)*. Three weed seed bank harvest depths and two grassland establishment rates across five timings were evaluated for preventing weed re-infestation. Prior to fraise mowing, soil cores were collected from the study area and subjected to seedbank analysis. Similarly, seedbank analysis of harvested fraise mow debris was completed to determine efficacy of weed seed harvest. Despite a reduction in the smooth crabgrass weed seed bank and a significant effect of timing and harvest depth on re-infestation, smooth crabgrass surpassed thresholds (<15%) by the end of the first full growing season, indicating the need for additional follow-up treatments for commercially acceptable control.

STACKED WEED MANAGEMENT STRATEGIES ARE EFFECTIVE AND CAN INCREASE SOIL QUALITY DURING ESTABLISHMENT YEARS OF AN ORGANIC, HIGH-DENSITY APPLE ORCHARD IN NEW YORK. K. Brown*, D. Zakalik, M. Brown, and G. Peck, Cornell University, Ithaca, NY

ABSTRACT

Adoption of certified organic apple production in the northeastern United States is constrained in part by frequent summer precipitation that fosters weed germination and growth, including many persistent perennial weeds. Organic growers in this region have few options for effective weed management and often cite weeds as the most constraining management barrier to high density production. While previous research studies have shown mulch can be an effective weed control practice, organic growers currently favor mowing with occasional hand weeding. Developing an integrated approach to organic weed management could lead to increased adoption of organic production by reducing pesticide inputs, as well as labor and equipment costs. In 2015, a NOFA-NY certified organically managed experimental orchard with 'Honeycrisp'/'Budagovsky.9' trees trained as a tall spindle was established at Cornell Orchards in Ithaca, New York. In 2016, a no-intervention control, cultivation, and surface-applied wood chip mulch were implemented as main treatments. Sub-treatments of mowing and two organically approved herbicides were then overlaid to complete a randomized, split-plot design with four complete blocks. As part of this long-term, systems-level experiment, weed cover and biomass, weed species biodiversity, soil properties, foliar nutrition, and tree growth are measured annually. In 2017, plots with only wood chip mulch had 29% and 38% less weed biomass compared to stand-alone cultivation and control plots, respectively. Across all treatments, weed biomass increased each growing season and, in 2019, weed biomass was not different among the stand-alone main treatment plots. Among the sub-treatments, weed biomass has always been higher in the control plots compared to herbicide and mowing, which were never different from one another. Improvements to soil health have been constant in the mulched plots, which currently have 27% more soil organic matter than cultivation and control plots. Mulched plots also have the highest values for active soil carbon, aggregate stability, soil respiration, and ace soil protein index. Early benefits for weed management and soil health under wood chip mulch have failed to enhance tree growth; cultivated trees have grown 43% more than control trees and 46% more than mulched trees since spring of 2016. While our study found that integrating mulch into organic apple orchards can reduce early weed competition and improve soil health, these benefits came at the cost of tree growth – possibly due to water saturated soils or nitrogen immobilization under the mulch. Understanding the best management practices for weed control in establishing organic orchards will require observations over a number of years under different soil types and varying weather patterns.

WHAT'S NEW? S.A. Mathew*, P. Berry, D. Black, D. Bowers, and J. Weems, Syngenta, Gaithersburg, MD

ABSTRACT

Tavium® Plus VaporGrip® Technology herbicide is a premix of *S*-metolachlor and dicamba together with VaporGrip® Technology. This combination product received EPA registration in 2019 and provides consistent control of emerged broadleaf weeds and excellent pre-emergence control of grasses and small seeded broadleaves like Palmer amaranth in Roundup Ready 2 Xtend® Soybeans or Bollgard II® XtendFlex® Cotton.

Coming soon is Acuron® GT is a premix of *S*-metolachlor, mesotrione, bicylclopyrone, and glyphosate for postemergence use in glyphosate tolerant corn. Upon registration, Acuron GT will provide improved postemergence and residual control of broadleaf weeds while maintaining the high level of crop safety experienced with Syngenta's Halex GT® herbicide.

Gramoxone® SL 3.0 is a new formulation of paraquat containing 3.0 lbs. of paraquat active ingredient per gallon. This formulation provides improved cold tolerance over Gramoxone SL 2.0 and containers less than 120 gallons will be sold in SecuraLink™ closed-system packaging for improved handling stewardship. Gramoxone SL 3.0 will be available for use in all burndown, post-directed, and harvest aid applications currently available in Syngenta's paraquat formulation Gramoxone SL 2.0.

To provide greater residual control and flexibility in soybean weed management programs, Syngenta has submitted changes to the Dual Magnum® and Dual II Magnum® labels to allow for increases in the maximum seasonal use rates of *S*-metolachlor in soybeans. Upon EPA approval, these changes will also increase the maximum post-emergence single application use rate and reduce the pre-harvest interval from 90 days to 75 days after application. Syngenta's other *S*-metolachlor containing products used in soybean will also align with these changes.

Acuron® GT is not yet registered for sale or use in the United States and is not being offered for sale. Tavium Plus VaporGrip Technology is a Restricted Use Pesticide.

INTRODUCTION AND OVERVIEW OF MON 301107: A NEW GLYPHOSATE FORMULATION.

D.J. Mayonado*, R.A. Recker, C.M. Mayo, R.F. Montgomery, and G.G. Camargo, Bayer, Hebron, MD

ABSTRACT

MON 301107 is a new glyphosate formulation. Field trials conducted in 2017 and 2018 in 54 locations evaluated MON 301107 for postemergence weed control compared to commercial standards. The experimental design was a split-plot arrangement with 3-4 replications. Whole-plots consisted of different glyphosate rates and the sub-plots were various glyphosate formulations. Results from 2017-18 trials, 14 days after treatment, indicated MON 301107 at 1120 g a.e. ha^{-1} provided broadleaf control and grass control that was not statistically different than commercial standards at 1120 g a.e. ha^{-1}. Additional field trials conducted in 2019 evaluated crop safety of MON 301107 compared to Roundup PowerMAX® herbicide when used postemergence on multiple crops. The experimental design was a split-plot arrangement with 3-4 replications. Whole-plots consisted of different herbicide treatments and the sub-plots had either MON 301107 or Roundup PowerMAX as the glyphosate formulation utilized in the herbicide treatment. Glyphosate formulation was not a significant treatment factor for any of the percent injury evaluations for field corn hybrids with Roundup Ready® 2 Technology, Soybean with Roundup Ready 2 Yield® Technology, Roundup Ready® Alfalfa, or Roundup Ready® Sugarbeet. These results demonstrate MON 301107 can provide non-selective foliar control of both grass and broadleaf weeds and has a comparable crop safety profile to Roundup PowerMAX.

THE BICYCLOPYRONE WEED CONTROL ADVANTAGE IN A NEW PREMIX PRODUCT CONCEPT FOR CORN. E. Hitchner*, R. Lins, S. Cully, T. Beckett, and G. Vail, Syngenta, Greensboro, NC

ABSTRACT

Acuron® GT is a new herbicide coming soon from Syngenta for weed control in glyphosate tolerant field corn. Acuron GT will contain *S*-metolachlor, mesotrione, bicyclopyrone and glyphosate for postemergence application with knockdown and residual control of grasses and broadleaves. In 2019, field trials were conducted to evaluate Acuron GT for weed control and crop tolerance. Results show that Acuron GT effectively controls many difficult weeds and provides improved residual control and consistency compared to other commercial standards. Acuron GT is not registered for sale or use in the US and is not being offered for sale.

MANAGEMENT OF VOLUNTEER CORN WITH FLUAZIFOP PLUS DICAMBA TANK MIXES.

S.A. Mathew*, P. Eure, M. Hay, and E. Parker, Syngenta, Gaithersburg, MD

ABSTRACT

Volunteer corn can significantly reduce soybean yield if left uncontrolled. Soybean yield loss caused by volunteer corn is dependent on the density and duration of interference. Management of volunteer corn in soybean includes reducing corn grain losses at harvest, tillage, cultivation, and chemical control. Use of Group 1 herbicides in soybeans to control glyphosate tolerant volunteer corn is common.

Commercialization of dicamba tolerant soybean (*Glycine max* L. (Merr.)) increased the frequency of dicamba and Group 1 herbicide tank mixtures for control of glyphosate tolerant volunteer corn (*Zea mays* L.). Previous research documented the antagonism of monocot control when dicamba is added to Group 1 herbicides. Therefore, the objectives of this research were to (1) Evaluate Fusilade® DX (fluazifop-P-butyl) herbicide volunteer corn control with and without dicamba (2) Discuss factors that influence volunteer corn control when using Fusilade DX + dicamba tank mixtures.

This research concluded that Fusilade DX is an effective tool for managing volunteer corn in soybeans when tank mixed with and without dicamba. Volunteer corn management may also be influenced by corn height, tank mix partners, and adjuvants.

STILL SEARCHING FOR A SOYBEAN BURNDOWN PROGRAM. M.J. VanGessel*, Q. Johnson, and B. Scott, University of Delaware, Georgetown, DE

ABSTRACT

A critical step in successful weed control programs is to “start clean”, with no living weeds present at time of planting. This is been a challenge for no-till soybeans that are often planting in early May in Delaware. Most fields have required two herbicide applications prior to planting soybeans. Three trials were initiated in the spring of 2019 to evaluate herbicide programs to provide advice on weed control prior to planting no-till soybeans. The first trial was designed to evaluate various herbicide options for control of existing winter annual weeds. All treatments included glyphosate, with tankmix partners including halauxifen, 2,4-D, dicamba, saflufenacil, or glufosinate. Horseweed control was best with tankmixes of dicamba or saflufenacil; while cutleaf evening primrose control was best with tankmixes of glufosinate, saflufenacil, or 1.1 kg ae/ha of 2,4-D.

Early preplant (EPP) treatments are encouraged in order to treat winter annuals while they are small and most susceptible. However, that means there is opportunity for winter annual and/or summer annual weeds to emerge between the time of application and planting. So inexpensive residual herbicides were evaluated for use during this pre-plant period. Metribuzin, metribuzin plus chlorimuron (Canopy), rimsulfuron plus thifensulfuron (Basis Blend), flumioxazin, saflufenacil or pendimethalin were tankmixed with glyphosate plus glufosinate. Only flumioxazin provided more than 75% control of Palmer amaranth (87%); ivyleaf morningglory control was 99 and 85% for Canopy and flumioxazin, respectively, while all other treatments provided less than 40% control. In addition, Canopy, Basis Blend, and flumioxazin provided the best large crabgrass control with 82 to 91% control.

Metribuzin, either as a premix such as Canopy, or alone is a cost-effective herbicide for no-till. It not only provides good control of many common species, it also improves the performance of paraquat or glufosinate when used under less than favorable conditions. However, growers have concern about crop safety when metribuzin in used prior to planting soybeans. A trial examining metribuzin rates for preplant and at planting on soybean safety was conducted at two locations. Early preplant rates of metribuzin included 0, 205, 130, or 211 g/ha and rates at planting were 130, 211, or 235. All combinations were included in the trial. No soybean injury was observed and yields did not differ among the treatments.

Regardless of the residual herbicides used, two herbicide applications are going to be the most consistent to control winter annual weeds and provide a “clean” field at planting. Based on one year of results, metribuzin could be used as part of the EPP application and be applied at planting without soybean injury.

BURCUCUMBER IN CORN AND SOYBEAN: WHICH HERBICIDES ARE EFFECTIVE? D. Lingenfelter* and J. Wallace, Penn State, University Park, PA

ABSTRACT

In the late 1990s, researchers at Penn State University conducted various experiments on burcucumber (*Sicyos angulatus*) biology and management. From these studies, certain herbicide options and tactics were recommended to farmers to help manage this problem weed in agronomic crops. Since then, not only does this weed continue to cause problems in crops but also, a few new herbicide active ingredients that might be effective on burcucumber have come to the market. Thus, no-till corn and soybean field studies were established in 2018 and 2019 in Pennsylvania (Landisville, Lancaster Co.) to evaluate PRE and POST herbicide programs containing these newer active ingredients as well as programs that were originally recommended. Studies were arranged in a randomized complete block design with three replications. Herbicides were applied with a small-plot, CO2-backpack sprayer system that delivered 15 GPA. In the corn trials, PRE herbicides were applied on late April/early May and POST about 30 days later when burcucumber seedlings were 1 to 4 inches tall. In the soybean study (2018 only), a broadcast PRE application (*S*-metolachlor + metribuzin) was applied on May 15 and the POST treatments on June 20 when burcucumber was 2 to 18 inches tall (9-inch average height). Corn treatments included various combinations of atrazine, simazine, bicyclopyrone, bromoxynil, dicamba, fluthiacet, glufosinate, glyphosate, isoxaflutole, mesotrione, *S*-metolachlor, primisulfuron, prosulfuron, and tembotrione. Primary POST soybean treatments included: chlorimuron, dicamba, fomesafen, glyphosate, imazethapyr, lactofen, and thifensulfuron. (Some of these combinations were used as premix formulations.) All the spray mixtures contained the necessary adjuvants. Visual weed control ratings were taken periodically throughout the growing season. In the corn studies, all of the PRE treatments provided <90% burcucumber control prior to the POST application. Late season (September) ratings showed that POST treatments containing prosulfuron or tembotrione provided 83-92% burcucumber control. The other POST treatments ranged from 55-80% control. In the soybean study, POST treatments containing chlorimuron, dicamba, and glyphosate provided 86-95% burcucumber control in August; while the others provided less control (73-84%). These studies of newer herbicide options (e.g., bicyclopyrone, fluthiacet, glufosinate, mesotrione, etc.) suggests that these herbicides are not necessarily better than previously recommended products. Products that provide residual control such as atrazine, chlorimuron, and prosulfuron continue to be effective herbicides for burcucumber management. However, herbicide application timing and rate will likely affect efficacy to a certain level as well. Also, with the advent of dicamba-resistant soybean varieties, over-the-top applications of dicamba (especially if tank-mixed with chlorimuron) in soybean can help to improve control of burcucumber in this setting. The same principles of burcucumber management still apply today just as they did 20 years ago. The key points to burcucumber management include: i) avoid spreading the seed via harvest or tillage equipment; ii) use no-till practices to keep the seeds on the soil surface, thus allowing germination to occur over a shorter time period and reducing the number of germination flushes (this also improves herbicide effectiveness and performance); iii) plant shorter season corn varieties to allow earlier silage harvest (ensiling kills green/immature burcucumber seed); iv) use two-pass programs that include effective foliar and residual herbicides are required for season-long control; and v) pre-harvest burcucumber control is usually not effective. As newer herbicides or technologies come to the market, additional research will be necessary to determine their value in managing burcucumber.

COMPARING MODELS FOR WEED SEEDLING EMERGENCE IN THE NORTHEAST. C. Marschner*, A. Degaetano, C. Sousa-Ortega, M.J. VanGessel, and A. DiTommaso, Cornell University, Ithaca, NY, NY

ABSTRACT

A multistate group of weed ecologists from Maine to Virginia are collaborating to collect weed emergence data on summer annuals, to support the development of a decision support tool for farmers to predict weed emergence and improve weed management timing. The study was initiated in 2019, with two more years of data collection planned. The study is focused on common lambsquarters (*Chenopodium album* L.), large crabgrass (*Digitaria sanguinalis* (L.) Scop.), smooth pigweed (*Amaranthus hybridus* L.), morning-glories (*Ipomoea* spp.), and common ragweed (*Ambrosia artemisiifolia* L.), but individual researchers are collecting more data on problem species for their states. Collaborators from California, Spain, and New York continued to compare and refine model options. Initial work with equations from two existing sets of models produced good but variable results by species when applied to preliminary data collected in New York and Delaware. A set of models for the five target species are being developed to compare with the existing models. Next steps will be to collect a second year of data and select final models for the focus weeds, verify the soil moisture model to be used in the online tool, and begin the online decision support tool development, incorporating soil type information into the predictive model used by the tool. The final product of this research will be housed on the Northeast Environment and Weather Applications (NEWA) website.

HSD MILL EFFICACY ON THE WEED SEEDS OF MIDWESTERN AND MID-ATLANTIC UNITED STATES. L.S. Shergill*, K. Bejleri, A. Davis, and S. Mirsky, USDA-ARS & University of Delaware, Beltsville, MD

ABSTRACT

The Harrington Seed Destructor (HSD), a harvest weed seed control (HWSC) technology that destroys weed seeds in seed-bearing chaff material during grain crop harvest, has been highly effective in Australian cropping systems. However, the HSD has never been tested on weeds common to soybean production systems in the Midwest and Mid-Atlantic US. We conducted stationary HSD testing and winter burial studies during 2015-2016 and 2017-2018 to determine (i) the efficacy of the HSD to target weed seeds of seven common weeds in Midwestern and five in Mid-Atlantic US, and (ii) the fate of HSD-processed weed seeds after winter burial. The HSD destroyed 93.5-99.8% of weed seeds in 2015 and 85.6-100% of weed seeds in 2017. The weak relationships (positive or negative) between seed size and seed destruction by HSD, and high percentage of weed seed destruction by HSD across all seed sizes indicate that the biological or practical effect of seed size is limited. The HSD-processed weed seeds that retained at least 50% of their original size, labeled as potentially viable seed (PVS), were buried for 90 d over winter to determine the fate of weed seeds after winter burial. At 90 d after burial (DAB), the HSD-processed PVS were significantly less viable than unprocessed control seeds, indicating that HSD processing physically damaged the PVS and promoted seed mortality over winter. A very small fraction (< 0.4%) seed of the total weed seed processed by HSD remained viable after winter burial. The results presented here demonstrate that the HSD is highly effective in increasing seed mortality and could potentially be used as a HWSC tactic for weed management in this region.

BIOLOGY OF HORSEWEED (*CONYZA CANADENSIS*): CHARACTERISTICS AND SIGNIFICANCE. P.C. Bhowmik* and D.O. Soares, University of Massachusetts, Amherst, MA

ABSTRACT

Conyza canadensis (L.) Cronq. is known as horseweed or mare's-tail. This species is known as *Erigeron canadensis* L. It is a native winter or summer annual. The plants are erect with one to several stems reaching 30 to 150 cm tall. Stems are typically unbranched at the base unless damage has occurred to the apical growing points. The leaves are linear to oblanceolate, 2 to 8 cm long and 2 to 8 mm wide. The leaf margins are serrated. The inflorescence is a loose panicle. The numerous flower heads are very small, 2 to 4 mm tall and 3 to 7 mm wide. The rays are white or purplish and are very small, only reaching 0.5 to 1.0 mm in length. The fruit is an achene with a white bristly pappus. Horseweed plants produced as many as 200,000 seeds per plant in a no-tillage site in Massachusetts. There were approximately 700,000 seeds per pound.

Horseweed is native throughout much of North America. It was introduced into Europe in the middle of 17th Century. Horseweed is a weed of more than 40 crops according to Holm *et al.* (1997). Horseweed is common in roadsides often found in newly abandoned fields and waste areas. Changes in tillage practices often result in changes in the weed species composition of conventional or conservation tillage systems in agricultural crops. Now, this species has become a common weed in agricultural cropping systems, especially in no-till or minimum tillage systems. In North America, horseweed infests orchards, vineyards, nursery crops, field crops such as corn, soybean and cotton, pastures and rangeland.

Management of this species in agricultural cropping systems poses challenges primarily because of the evolution of resistance. This species has developed resistance to several herbicides. In 1982, it was first reported as resistant to simazine in orchards in United Kingdom. The species has been reported as resistant to paraquat in Mississippi (1994) and resistant to glyphosate in Delaware (2000), and to other herbicides in recent years. This species has become a significant weed species in agricultural crops, Christmas tree and nursery crop productions. We must keep an eye on this species as we develop alternative management options.

COMPARING HERBICIDE APPROACHES FOR DIFFERENT COVER CROP TERMINATION TIMINGS. M.J. VanGessel*, Q. Johnson, and B. Scott, University of Delaware, Georgetown, DE

ABSTRACT

Cereal rye is used widely in the Mid-Atlantic region to address nutrient management and water quality in the Chesapeake Bay Watershed. Delaying the cereal rye termination can expand the benefits (or ecological services) that rye can provide. Previous research at University of Delaware has shown that delaying cereal rye termination until 2 wks prior to planting improved overall weed control compared to termination 4 wks prior to planting. This trial was established in the fall of 2018 and soybeans planted in 2019 to determine if late terminated cereal rye can reduce herbicide inputs, with either fewer products used and or fewer herbicide applications. This was a two-factor trial with cereal rye termination timing as one factor and herbicide approach as the second factor. Cereal rye was drilled in the fall and terminated at 4 wks or 2 wks early preplant (EPP) or 1 day after planting (planting green). Glyphosate plus 2,4-D was used to terminate the cereal rye. The six herbicide approaches were: 1) non-selective only (burndown or BD); 2) BD plus a residual herbicide; 3) BD plus residual herbicide followed by (fb) POST application of glufosinate; 4) BD plus residual herbicide fb POST application of glufosinate plus residual herbicides; 5) BD fb residuals applied at planting fb POST application of glufosinate plus residual herbicide; or 6) BD fb early-POST (EPOST) plus residual herbicides.

Cereal rye biomass at time of termination ranged from 1100 kg/ha to 3560 kg/ah for the 4 wks EPP and 1 day after planting, respectively. Excellent horseweed and cutleaf evening primrose control was observed with all plots treated with glyphosate plus 2,4-D. Palmer amaranth control was less than 90% for all herbicide approaches used with early cereal rye termination. Including the residual herbicide with the BD application was not as consistent for Palmer amaranth control as delaying the residual application until planting fb a POST application or using an EPOST application. Morningglory control was best with 2 wks EPP or terminating 1 day after planting with a residual herbicide fb by a POST with residual, or BD fb residuals at planting or using an EPOST approach, but no treatment provided better than 88% control. Main effects were significant for yield with planting green and rye termination 2 wks EPP yielding higher than rye terminated 4 wks EPP; and yields were higher with use of residuals at planting, using an EPOST approach, or using BD fb POST with residual herbicide.

More research is needed to ensure the consistency of these results, but preliminary results indicate herbicide intensity can be reduced when BD application is made at planting without reducing weed control or soybean yield.

RELATIVE SENSITIVITY OF COVER CROPS TO SEEDLING ROOT AND SHOOT INHIBITORS.

J.M. Wallace* and E. Maloney, Pennsylvania State University, University Park, PA

ABSTRACT

Integrating cover crops in no-till crop production systems increases crop diversity, decreases soil erosion, and increases soil biological health. To maximize these ecosystem services, growers are prioritizing cover crop seeding directly after crop harvest, experimenting with cover crop interseeding in corn and soybean, and utilizing diverse cover crop mixtures. Greater understanding of the potential for residual herbicides to injure or reduce establishment rates of cover crops is needed to consistently realize conservation goals. In support of recent field trials, we conducted a series of greenhouse dose response assays that evaluated the response of twelve cover crop species to commonly used seedling shoot (15) and seedling root (3) inhibiting herbicides. Three experiments (grass, legume, non-legume cover crops) were conducted using a three-factor RCB design with five replicates and two experimental runs. Experimental factors included four cover crop species and five herbicides (acetochlor, dimethenamid, *S*-metolachlor, pyroxasulfone and pendimethalin) applied at seven rates standardized based on label rates (1X, 1/2 X, 1/4 X, 1/8 X, 1/16X, 1/32X, 0X) for medium-textured soils in corn. Cover crop species included annual ryegrass (Kodiak), cereal rye (VNS), triticale (Elevator), sorghum sudangrass (AS5201), crimson clover (Dixie), medium red clover (VNS), hairy vetch (Purple Bounty), Austrian winter pea, winter canola (Wichita), Daikon radish, forage rape (Dwarf Essex), and buckwheat (Lifago). Experiment units (pots) were filled with 2:1 sand to potting mix medium and sown with five cover crop species prior to application of herbicides at 15 gpa in a track sprayer. Aboveground biomass was harvested 28 DAT, dried and weighed. Four parameter log-logistic models were fit for each herbicide by cover crop combination and the ED50 parameter estimate was used to compare relative sensitivities. Annual ryegrass and sorghum sudangrass were more sensitive across herbicides compared to triticale and cereal rye, which were tolerant to pyroxasulfone and pendimethalin at 1X rates. Crimson clover was the most sensitive legume species across herbicide treatments and acetochlor was the most injurious herbicide across legume species, with mean ED50 values below 1/8X product rates. Brassica cover crops (canola, rape, radish) were tolerant at 1X rates to pendimethalin, *S*-metolachlor and dimethenamid but were highly sensitive to pyroxasulfone. Buckwheat was tolerant to pendimethalin but highly sensitive to seedling shoot inhibiting herbicides. When coupled with field-level observations, these results can be utilized to generate rules-of-thumb and general guidelines for integrating diverse cover crop mixtures into annual crop rotations, or conversely, to select herbicide programs that limit impact on cover cropping programs. Future work will expand this methodology to additional herbicide sites-of-action and cover crop species.

RESIDUAL HERBICIDE PROGRAM EFFECT ON INTERSEEDED ANNUAL RYEGRASS IN CORN GRAIN. T.R. Mazzone* and J.M. Wallace, The Pennsylvania State University, State College, PA

ABSTRACT

Establishing cover crops post grain harvest can be challenging for growers in areas of the mid-Atlantic region. Interseeding cover crops within the growing season can improve cover crop establishment, but also increases risk of exposure and injury to residual herbicide programs. We conducted two field trials to assess interseeded annual ryegrass (KB Supreme) response to 10 residual corn herbicide programs. Field trials were conducted using a RCB design with four replications. Trial locations were RELARC, Pennsylvania Furnace, Centre County, PA and SEAREC, Manheim, Lancaster County, PA. At corn planting, residual herbicide treatments were applied with a glyphosate burndown. Residual herbicide treatments were applied at a 1X label rate and included rimsulfuron + thifensulfuron (Basis Blend), saflufenacil +a dimethenamid-P (Verdict), pyroxasulfone + fluthiacet (Anthem Maxx) + atrazine, isoxaflutole (Balance Flexx) + atrazine, thiencarbazone + isoxaflutole (Corvus) + atrazine, pendimethalin (Prowl H20) + atrazine, mesotrione (Callisto) + atrazine, acetochlor + mesotrione (Harness Maxx) + atrazine, metolachlor + atrazine + mesotrione (Lexar EZ), and metolachlor + atrazine + mesotrione + bicyclopyrone (Acuron). A glyphosate burndown only treatment and a glyphosate burndown followed by glyphosate post-emergence were included as controls. Annual ryegrass was interseeded at 4-6 weeks after corn planting or V3-V4 corn growth stage. Cover crop and weed aboveground biomass was harvested at approximately 84 and 140 days after herbicide application, oven-dried, and weighed. Analysis of variance was used to assess treatment differences using the *lmer* package in R.

At both RELARC and SEAREC sites, residual herbicide programs did not significantly affect annual ryegrass biomass 84 days after treatment. At RELARC, when compared to the untreated check, Anthem Maxx and Corvus reduced weed biomass 84 days after application. Acuron, Lexar and Prowl reduced weed biomass at both 84 and 140 days after application. Callisto treatment reduced weed biomass at 140 days after application. At SEAREC, Callisto treatment had the highest numerical weed biomass of all treatments, however the weeds were composed of 95% grass versus broadleaf species. In other treatments weed biomass was not affected by residual herbicide programs when compared to the control. Herbicide degradation rates in soil are a function of weather conditions, soil texture and herbicide properties. Therefore, the absence of annual ryegrass injury observed in these field trials is likely a function of environmental conditions that promoted herbicide dissipation early in the corn growing season. Our preliminary data suggests that annual ryegrass interseeded at V3 can establish even in the presence of full-rate, residual herbicide programs. However, evaluation of these treatments over a range of environmental and soil conditions is needed to support further development of management recommendations.

COMPARISON OF ROLLED VERSUS STANDING COVER CROP RESIDUE FOR COMMON RAGWEED (*AMBROSIA ARTEMISIIFOLIA*) SUPPRESSION. L.S. Rector, M.L. Flessner*, K.B. Pittman, D.S. McCall, and W.E. Thomason, Virginia Tech, Blacksburg, VA

ABSTRACT

Greater cover crop biomass has been associated with greater summer annual weed suppression, especially small seeded weeds that require light for germination. Both rolling cover crop residue and leaving residue standing are common practices, but little research exists comparing these residue management methods. Therefore, we evaluated the effects of cover crop residue biomass level (none, low, medium, or high), cover crop species (wheat or cereal rye), and cover crop termination method (roller crimper and herbicide or herbicide alone) on weed control and light penetrating the cover crop canopy. A greenhouse study was conducted to compare cereal rye biomass rate with common ragweed establishment and light interception. Additionally, a small plot field study was conducted further comparing factors of cover crop species, termination method, and biomass levels on weed control, light penetration, soil moisture, and soil temperature. Across greenhouse and field experiments, as cover crop biomass increased, weed control increased and light penetration decreased. Cover crop species had no effect on common ragweed control, but standing residue provided greater weed control than rolled residue up to 8400 kg ha^{-1} for the field study. Cereal rye intercepted more light compared to wheat in both studies, but in the field study, there was no difference in light interception after 6000 kg ha^{-1} of cover crop biomass. In the field study, rolled cover crop residue intercepted more light than standing residue, but differences never exceeded 10%. In the field study, increased cover crop biomass led to at least a 20% reduction in soil temperature and 40% increase in soil moisture at 10000 kg ha^{-1} of cover crop biomass compared to no cover crop. In the field study, rolled cereal rye provided a greater reduction in soil temperature than standing cereal rye, rolled wheat, or standing wheat. In the field study, wheat led to a greater increase in soil moisture compared to cereal rye. Producers with less than 8400 kg ha^{-1} biomass at termination should leave residue standing to the extent possible. Obtaining maximum cover crop biomass will lead to greater weed control through increased light interception to prevent weed establishment, reduced soil temperature to delay weed establishment, and increased soil moisture, which may favor weed establishment.

MAINE AND COLORADO POTATO GROWERS RESPONSES TO A CHANGING CLIMATE.

J. Jemison*, and Ruth Sexton, University of Maine, Orono, ME

ABSTRACT

Potatoes are important to the agricultural economy of Maine and the San Luis Valley of Colorado. Each region produces approximately 50,000 acres of potatoes for the processing, seed or tablestock industries. Understanding how growers perceive changing or variable weather is important to understanding how receptive they may be to climate-smart production measures. In 2019, interviews were conducted with 14 Maine and 12 Colorado potato growers related to irrigation and soil management. In these interviews, growers were asked three related questions: 1) Do you notice changes in the weather patterns affecting your potato production? If yes, please elaborate; 2) In the last decade, do you think you have seen more soil loss through runoff and or wind erosion or less? If less, why?; and 3) Do you think the growing seasons are warmer/drier than they used to be ? is it affecting how often and much you irrigate? These interviews were recorded, transcribed and analyzed with NVivo 12 software. Responses of Maine growers were also compared to those made in 2011 by a focus group of a dozen Maine potato growers to see how views have changed over the decade related to variable weather and soil and water management. In 2019, four Maine and three Colorado growers rejected the idea that increasingly variable weather affected their soil and water management. In each case, growers made statements that they perceived heavier rainfall happening now than previously (Maine) or that the growing season may be warmer and longer (Colorado), but they stated that increasingly variable weather did not influence their soil and water management. Another five Maine and three Colorado growers were uncertain if they thought variable weather had affected management. The remaining five Maine and five Colorado growers made definitive comments about how variable weather influenced their overall potato production and management. Of those growers rejecting whether the climate was changing, each Maine farmer related comments about increasingly variable rainfall, increasingly intense rain events which no one mentioned in 2011. Much of this decade, Maine farmers have seen a notable increase in the number of two inch rain events in a 48 hour period. Colorado farmers did not mention an increase in the amount of wind erosion, but most spoke with some concern about the lack of snowfall in the mountains, early snow melt not coinciding with their need for water, and hotter Junes leading to higher irrigation demand while water rates continue to climb. Growers advocating for increasingly variable weather spoke directly about their concern for a changing climate affecting their ability to produce. An example quote includes this: “It's (climate change) happening. We'd better figure out how to handle it. That goes right back to having the capability to irrigate if it's extraordinarily dry.” Funding to support more interactive discussions of how climate change can affect potato production might help convince growers to implement practices to help hold soil, irrigate in a timely manner while setting up drainage systems to get water off the fields when it is in excess.

GLYPHOSATE MOBILITY FROM TRANSPORT-PRONE MARGINAL SOILS. B.K. Richards*, S. Pacenka, and T.S. Steenhuis, Cornell University, Ithaca, NY

ABSTRACT

Recent environmental surveys report widespread detections of the herbicide glyphosate [N-(phosphonomethyl)glycine] in surface waters despite its strong immobilization and rapid biodegradation in soils. Since 2015 we have carried out high-frequency sampling campaigns (5 in spring, 1 in fall) following controlled spray applications on an experimental perennial grass field site with wetness-prone marginal soils. We have monitored dissolved glyphosate concentrations in the field outflow (runoff and shallow drainage) using liquid chromatography-mass spectrometry and ELISA immunosorbent assays. Rainfall-triggered outflow events began between 3- and 17-days following spray application. Outflow concentrations of dissolved glyphosate varied widely from nondetectable up to 100 $\mu g\ L^{-1}$, peaking during each significant outflow event essentially concurrently with the flow peak. Successive concentration peaks tend to become smaller regardless of the magnitude of subsequent flow events. Cumulative mass losses in outflow across the different campaigns have ranged from 0.06 to 1.0 percent of applied glyphosate.

Cumulative glyphosate losses in outflow were not associated with total rainfall during the post-spray sampling period. In most – but not all - cases, losses were better predicted by soil hydrologic conditions at the time of spraying as reflected by the 7-day cumulative pre-spray rainfall, with wetter antecedent conditions favoring greater cumulative mobilization, especially if time lags until mobilizing rain events were short. We are continuing to examine factors influencing mobilization.

CBD HEMP PRODUCTION: LESSONS LEARNED FROM COLORADO. J. Jemison*, University of Maine, Orono, ME

ABSTRACT

Hemp (*Cannabis sativa*, L.) is now legal in all 50 states due to the passage of the 2018 Farm Bill. Several states have had state-administered CBD cannabis production programs for a longer period of time, including both Maine and Colorado. On a sabbatical leave, I worked with one small scale (10-ac) and one large scale (500+-ac) grower to study production practices and learn about production issues to New England hemp growers. Both farms direct seeded and transplanted several different cannabis varieties. Due in part to a cool spring, germination rates on direct-seeded varieties was very poor, along the lines of 30% germination with a seed that cost approximately $1/seed. Colorado growers planted on much tighter density than can be done in the East in part to control weeds and to increase production, and as a result, Colorado growers were able to produce four times the amount of plants per acre than we can do in the Northeast. As well, black plastic was not used by either grower, and weed management was extremely problematic. It was dealt with by significant labor-intensive hand weeding. Weed management was by far the worst challenge in Colorado. The most important insect pests found in the San Luis Valley included piercing-sucking insects like aphids, leaf hoppers, and false cinch bugs. Diseases were not nearly the problem that they are in Northeast production systems. Center pivot irrigation systems allow growers to provide sufficient water for production, but the plants are generally dry within 20 - 30 minutes after irrigation. This leads to much more favorable conditions for growth. The majority of the labor savings comes with harvest as the cannabis can be combined and windrowed for drying instead of hand hanging plants to dry in barns. The economies of scale will make it extremely hard for Northeast US growers to compete with Colorado growers. In addition, the new proposed USDA hemp rules threaten all farmers' capacity to produce CBD hemp.

MANAGING HERBICIDE RESISTANT COMMON RAGWEED (*AMBROSIA ARTEMISIIFOLIA* L.) EMERGENCE AND GROWTH IN SOYBEAN. S.M. Hirsh*, M.J. VanGessel, and B. Beale, University of Maryland, Princess Anne, MD

ABSTRACT

Herbicide resistant common ragweed (*Ambrosia artemisiifolia* L.) is prevalent on the Lower Eastern Shore. In 2019, common ragweed populations were found to have two or three-way mode-of-action resistance on the Eastern Shore, and farmers have reported that herbicide-resistant ragweed prevalence is increasing. We investigated the effect of delayed cover crop burndown timing, and the effectiveness of residual herbicide products—Command (clomazone), Linex 4L (linuron), Dimetric (metribuzin), Command + Linex, Command + Dimetric, and Linex + Dimetric—on herbicide resistant common ragweed emergence and growth. We found that when cover crops were terminated April 4, there was higher common ragweed prevalence than when cover crops were terminated April 29 or at soybean planting (when herbicide was applied only once at cover crop burndown time). There were no differences in ragweed prevalence or soybean yield among the various residual herbicide product treatments. Also, there were no differences in soybean yield due to cover crop termination date; notably, planting soybean into standing cover crops did not reduce soybean yields.

EMERGING HERBICIDE RESISTANCE ISSUES IN NORTH CAROLINA. W.J. Everman*, E.A. Jones, and C. Cahoon, North Carolina State University, Raleigh, NC

ABSTRACT

[no abstract entered]

FACTORS DETERMINING APPLICATION ACCURACY WHEN USING A UAV-SPRAYER BASED ON PREVIOUS WEED MAPPING. J. Hunter III, T.W. Gannon, R.J. Richardson, J. Neal, and R.G. Leon*, NCSU, Raleigh, NC

ABSTRACT

Spray capable unmanned aerial vehicles (UAVs) are currently been marketed for conventional broadcast applications of liquid pesticides, but application time per flight is seriously limited by battery duration and payload capacity. Integrating UAV-based weed mapping with UAV-sprayers can reduce flight time, amount of herbicide spray, and maintain weed control. The present research indicated that the site-specific applications with a UAV-sprayer were more accurate targeting weedy areas compared with a ground-based broadcast application. The UAV system efficiency decreased as weed density and distribution increased. Nozzle type and flying speed were major drivers of application accuracy and drift risk with the UAV sprayer. The results indicated that the AIXR nozzle provided the most adequate and consistent coverage as application speeds increased. The HC and XR nozzles were highly susceptible to off-target movement, while the TTI nozzle did not provide acceptable coverage. Based on the results, integrating weed mapping and site-specific applications using UAV technology can become a powerful new tool for increasing efficiency and reducing cost in integrated weed management programs, but more research is needed to determine ways to reduce drift and off-target movement.

FERTILITY, TURF SPECIES, AND PLANT GROWTH REGULATOR EFFECTS ON FLORAL DENSITY IN A BULB LAWN. S.D. Askew*, M. Shock, J. Craft, and J. Brewer, Virginia Tech, Blacksburg, VA

ABSTRACT

Bulb meadows constitute a method to integrate managed turf areas so that they also provide spring floral displays that serve pollinator species. Three field trials and one greenhouse trial were conducted in 2018 and 2019, and suggest that bulb meadows can be successfully integrated into existing cool-season turf. In the fall of 2018, eight bulb types were randomly arranged in rows across 1-m wide plots and tall fescue or fine fescue sod was overlaid. Fertility programs for 2019 included no fertility, "turf fertility" consisting of 98 kg/ha nitrogen applied in fall, "bulb fertility" consisting of 146 kg/ha phosphorous and potassium and no nitrogen applied in early spring, "integrated turf and bulb fertility" consisting of bulb fertility and a reduced 49 kg/ha nitrogen in fall, and "integrated turf and bulb fertility plus turf suppressant" which also includes fluazifop at 105 g ai/ha applied in early spring. Additional field and greenhouse studies evaluated fluazifop and three other turf suppressants including glyphosate at 72 g ae/ha, metsulfuron at 4 g ai/ha, and trinexapac ethyl at 289 g ai/ha. Floral density was monitored via bloom counts every 3 days and showed differences in both periodicity and magnitude of floral density between species. Narcissus 'Baby Moon' had the highest bloom density in tall fescue at 45 blooms m^{-2} while Iris 'Katherine Hodgkin' and Narcissus 'Rip Van Winkle' had only 3 to 4 blooms m^{-2} in fine fescue. Fine fescue reduced bloom density of Iris, Crocus 'Cream Beauty', Crocus 'Ruby Giant', Muscari 'Venus', and both Narcissus varieties. Fertility effects will be measured in spring 2020. Metsulfuron and trinexapac ethyl were more injurious to bulb plants than glyphosate and fluazifop. Fluazifop reduced 4-wk turf clipping biomass by 87% and more than all other turf suppressants.

EFFECTS OF BASAL STEM GLYPHOSATE APPLICATIONS TO NURSERY TREES. A. Witcher*, Tennessee State University, McMinnville, TN

ABSTRACT

Glyphosate is the most common post-emerge herbicide for controlling weeds in field-grown nursery crops. Glyphosate is applied as a band spray to control weeds within the row, but the herbicide can contact tree stems when non-shielded spray tips are used. Glyphosate can be absorbed by tree bark and translocated within the plant, but the potential damage caused by these applications has not been extensively studied. The objective of this research was to evaluate basal glyphosate applications on growth of 'Summer Red' red maple (*Acer rubrum* L. 'Summer Red') over two growing seasons. One-gallon container-grown 'Summer Red' red maple plants received glyphosate treatments in 2018 and were then transplanted to 3 gal containers and subjected to the same treatments in 2019. Two experiments were conducted in summer 2019 at the Tennessee State University Nursery Research Center in McMinnville, TN. In both experiments, plants received basal (15 – 20 cm portion of lower stem) glyphosate treatments every 4 or 8 weeks (4 or 2 total applications, respectively) and a non-treated control was also included. In Experiment 1, plants received glyphosate at application volumes of 2, 20, or 40 gallons per acre (1 lb ai/A). In Experiment 2, plants were treated with glyphosate application rates of 1 or 2 lb ai/A. In both experiments, an ultra-low volume sprayer (Herbiflex 4) was also included for comparison with conventional sprayer applications. Trees received a total of 4 or 8 glyphosate applications over the two-year period. Shoot height was recorded at the beginning and end of the experiments, while shoot biomass was measured 4 months after the initial treatment.

At the end of 2019, no visual symptoms of herbicide damage were observed in either experiment. In Experiment 1, change in plant height and shoot biomass were not significantly affected by application volume or number of applications. In Experiment 2, final plant height and change in height were similar among all application rates and number of applications. Total shoot biomass was also similar among all treatments, although there was a 10% to 16% reduction in total shoot biomass in plants that received 8 glyphosate applications compared to the control. Although glyphosate can be absorbed through bark and translocated throughout the plant, basal glyphosate applications did not affect young red maple trees at the end of two growing seasons. Any visual or residual effects of basal glyphosate applications will likely take several years to materialize, thus further studies would be necessary to determine the long-term effects.

DOES WEED SUPPRESSION AND FERTILIZATION ENHANCE ESTABLISHMENT OF BAREROOT TREES IN A RIPARIAN BUFFER?
A.E. Gover*, E.N. Weaver, S.M. Jefferson, and E.B. Euker, Penn State, University Park, PA

ABSTRACT

[no abstract entered]

A PRELIMINARY STUDY OF THE NON-NATIVE VASCULAR FLORA OF FRESH KILLS PARK, STATEN ISLAND, NEW YORK. R. Stalter*, St Johns University, Queens, NY

ABSTRACT

The objective of this preliminary study was to document the non-native vascular plant species at 890 ha Fresh Kills Park, Staten Island, New York. The vegetation was initially sampled in September and October 2018 and monthly during the 2019 growing season. Additional collecting will be conducted during the 2020 and 2021 growing seasons. A preliminary list of the flora includes 299 species in 208 genera in 75 families. Families with the greatest number of invasive taxa are the Poaceae, Asteraceae, and Fabaceae. One hundred seven non-native vascular plant species compose 36% of the flora. Non-native taxa of greatest concern are *Phragmites australis*, *Celastrus orbiculatus*, and *Elaeagnus* spp. No eradication programs are currently in place to remove these taxa.

JAPANESE KNOTWEED AND MUGWORT MANAGEMENT WITH EARLY SPRING FOAMING STEAM APPLICATION. A. Senesac*, Cornell Cooperative Extension, Riverhead, NY

ABSTRACT

Recent developments in thermal weed control seem to offer an alternative to herbicides for managing weeds in amenity and certain natural areas. Two of the most troublesome invasive species in Northeastern US are mugwort (*Artemisia vulgaris*) and Japanese knotweed (*Reynoutria japonica*). Both spread laterally by rhizomes. However, the rhizomes and rootstock of the knotweed are highly condensed in the top few inches of soil and almost behave like the crown of a simple perennial. Japanese knotweed and mugwort are both deciduous and new shoots arise during the course of the spring season. This period is the most vulnerable time and offers the greatest opportunity for management of these weeds. The goal was to determine if the tender emerging shoots from two invasive rhizomatous weeds can be well controlled with exposure to thermal management supplied by Foamstream™. The Foamstream™ equipment combines hot water and foam, made from renewable plant oils and sugars. As the foam emerges from the applicator wand, it creates a layer of insulation that retards cooling and allows the destructive hot water to remain on the foliage longer, optimizing efficacy.

In April 2019, a site was identified at a nearby public preserve. The Japanese knotweed was growing along the side of a dirt roadway which allowed for the application vehicle to drive close by. The knotweed populations were divided into equal sized plots of 10' x 5' while the mugwort plots were 10'x 3'. The treatments were arranged in a Randomized Complete Block design and replicated four times. Following treatment on April 30th, the plots were evaluated for stem number, plant vigor and stem height regularly over the course of following six weeks. The treatments consisted of the mowed (control) and two periods of exposure of the shoots. One period (1X) simulated the standard exposure time for managing emerged weed species. The length of time that it took to cover the plot with foam was timed and that figure was used to determine the second period, which was 1.5-2 X the standard exposure period. The data were subjected to statistical analysis to determine if significant differences exist between the treatments. The same procedures were followed for the mugwort plots which were located in another area of the preserve.

The results suggest that re-emerging stem number, vigor and height were significantly reduced for the first 2-3 weeks following either Foamstream treatment compared to the mowed control. However, after this period, the Foamstream treated plots began to resume growth on a pace with the mowed plots. Most of the treatment effect was no longer visible by the end of the study at 6 weeks. Reapplication of the Foamstream at three-week intervals for the remainder of the season most likely would have caused a significant reduction in the underground reserves of either weed. In areas where repeated mowing as well as chemical herbicides are not allowed or practical, Foamstream may offer a legitimate alternative. However, if a large area is to be treated, there would be a necessity for nearby running water to refill the tank. This may be a limiting factor in the practicality of this method. But in amenity areas and hardscapes where water is available, the Foamstream is a potential new tool for landscapes and property managers.

JAPANESE STILTGRASS (*MICROSTEGIUM VIMINEUM*) GERMINATION PATTERN AND ITS IMPACT ON CONTROL STRATEGIES. J. Derr* and G. Payne, Virginia Tech, Virginia Beach, VA

ABSTRACT

Trials were conducted in southeastern Virginia to determine the germination period for Japanese stiltgrass and its impact on timing of herbicide applications. Germination declined from April to June, with no germination seen in mid-June. Applications of prodiamine reduced Japanese stiltgrass stand from 30 to 94%, depending on rate and timing, but even in the most effective treatment there was still at least 19% cover by August. Split prodiamine treatments that utilized May applications were more effective for crabgrass (Digitaria spp.) control than for Japanese stiltgrass, supporting the earlier germination pattern for Japanese stiltgrass. Optimum Japanese stiltgrass control occurred with a single March or a split December plus March application of prodiamine. Glyphosate applications in early May significantly reduced Japanese stiltgrass cover that month, but later germination resulted in approximately 25% cover in October, with 78% cover seen in untreated plots. No Japanese stiltgrass was observed in October when glyphosate was applied in either the third week of June or the third week of July. A single application of glyphosate is effective for Japanese stiltgrass control if applications are made after new germination has ceased in June.

EVALUATION OF HERBICIDES FOR CONTROL OF HONEYSUCKLE. J.C. Jodon, Jr.*, J. Plummer, D. Despot, and J. Sellmer, Penn State University, University Park, PA

ABSTRACT

Exotic shrub honeysuckle including Morrow's (*Lonicera morrowii A. Gray*), tatarian (*Lonicera tatarica L.*), and amur (*Lonicera maackii (Rupr.) Herder*) have become widespread along Pennsylvania roadsides disrupting vehicle sight lines, interfering with maintenance, and creating visibility hazards where wildlife cross. Past experiments demonstrated that 2,4-D, 2,4-D + triclopyr, and glyphosate were effective against honeysuckle; however, due to constraints of the standard CO_2 plot sprayer the applications were limited to side trimming the shrubs, which mimicked the approach used by roadside vegetation managers. Past injury and control ratings were based on the stems treated within the application pattern. Regrowth suggested that side trimming may not be an effective application method for honeysuckle control. In a continuing effort to find an effective control strategy, this experiment evaluated full coverage by six herbicide treatments: 4257 g 2,4-D ae/ha; 3361 g glyphosate ae/ha; 1681 g triclopyr ae/ha; 3361 g triclopyr ae/ha; 10084 g triclopyr ae/ha; 1681 g triclopyr ae/ha + 4257 g 2,4-D ae/ha; and an untreated check. The experiment was established as a randomized complete design with nine plants per treatment. Individual shrubs were measured to determine the area of each plant and herbicide applications were based on calculated canopy area. All herbicide treatments were applied at a carrier volume of 467 L/ha and included methylated seed oil at 1% v/v. Treatments were applied using a CO_2-powered sprayer equipped with a handgun and a PPX6 nozzle at 30 psi. The honeysuckle was treated on July 7, 2018. Treatments were visually rated for percent injury and control using the following rating system 0 = no injury–100 = complete necrosis on August 8, September 11, 2018, and July 10, 2019; 30, 64, and 366 days after treatment (DAT), respectively. All data were subject to analysis of variance and when treatment F-tests were significant ($p \leq 0.05$), treatment means were compared using Tukey's HSD separation test. After the first growing season, glyphosate at 3361 g ae/ha, 2,4-D at 4257 g ae/ha, and triclopyr at 10084 g ae/ha provided a minimum of 99% injury of honeysuckle. The untreated shrubs showed 29.44 percent leaf spot injury confirmed as Alternaria leaf spot. By 366 DAT, the most effective were glyphosate at 3361 g ae/ha (100%), triclopyr at 10084 g ae/ha (95.56%), 2,4-D at 4257 g ae/ha (95.22%), and triclopyr at 1681 g ae/ha + 2,4-D at 4257 g ae/ha (95%). Triclopyr at 1681 g ae/ha and triclopyr at 3361 g ae/ha resulted in 72.78 and 78.33 percent control of honeysuckle, respectively. The untreated check continued to show signs of Alternaria leaf spot damage with 22.89 percent control. Glyphosate was the most effective treatment; however, as a non-selective herbicide it may create bare ground below or adjacent to targeted shrubs as a foliar application. For glyphosate to be effectively employed seeding a permanent low growing grass groundcover after treatment to prevent erosion and following up in future seasons with a selective herbicide for control of broadleaf weeds would be advised. Alternatively, two seasons of triclopyr treatments may be effective without requiring seeding a low growing grass groundcover. Further data collection will determine the long-term effectiveness of the treatments employing total plant coverage applications.

PRECISION HERBICIDE APPLICATIONS TO ROSETTES AND SIMILAR GROWTH FORMS. B.F. McDonnell*, National Park Service, Bushkill, PA

ABSTRACT

A need was identified for a precise, replicable, and discrete method to apply herbicide to individual weed targets growing in a generally circular habit, such as *Cirsium arvense* rosettes. The *Cirsium arvense* targets were scattered within a restoration area in which high value pollinator plants had been sown. The application plan was to apply 0.09 kg/ha (0.078 lb/ac) aminopyralid to the rosettes while minimizing off target mortality. The method developed was to use a backpack sprayer fitted with a metered output spray gun, and a full cone spray tip, and is applicable to other weeds growing in a similar habit, such as a sedge. Using a TG-3 or TG-4 spray tip, a MeterJet spray gun calibrated to apply a one ml dose, and held at a height (305 mm, 1 ft) to produce a 200 to 250 mm (8-10") diameter spray pattern, the spray equipment is calibrated to apply an approximate rate of 24 GPA. The specific equipment, application method, and calibration procedure will be presented.

BROADLEAF WEED CONTROL ALTERNATIVES IN COOL-SEASON TURFGRASS. P.C. Bhowmik*, University of Massachusetts, Amherst, MA

ABSTRACT

Several field studies were conducted to evaluate selective control of various broadleaf weeds in our experimental site. Common weeds included common dandelion (*Taraxcum officinale*), white clover (*Trifolium repens*), broadleaf plantain (*Plantago major*), buckhorn plantain (*Plantago lanceolate*), yellow woodsorrel (*Oxalis st*ricta), heal all (*Prunella vulgaris*), common chickweed (*Stellaria media*), and other species. Experiments were conducted in 2017 to 2019. Experimental areas were mixed turfgrass primarily of Kentucky bluegrass. Experimental plots were maintained at a 1.5 inch cutting height and the clippings were left on the plots. The area was fertilized with 0.5 lb/N twice a year. All treatments were applied to 3.5 by 10 feet plots with a CO_2-backpack sprayer at a pressure of 22 psi in 50 gpa. POST treatments were applied on June 26, 2017, June 19, 2018 and June 17, 2019. Turfgrass injury was visually estimated on a scale of 0 to 100% (0%=no injury and 100%=dead turfgrass) and turfgrass density was rated on scale of 1 to 9 (where 1=thin stand and 9=dense stand). Weed control was visually estimated on a scale of 0 to 100% (where 0%=no weed control and 100%=complete control) 2, 4, 8 and 12 weeks after treatment (WAT).

Arylex (halauxifen-methyl) is a new class of synthetic auxin (GR 4). It belongs to arylpicolinate class and it is a systemic POST herbicide with reduced volatility. Several new products containing this new active ingredient were evaluated. Several new combinations such as GF-3566 (halauxifen-methyl + fluroxypyr + 2,4-D choline), GF-2687 (halauxifen-methyl and florasulam), Switchblade (halauxifen-methyl + fluroxypyr + dicamba) were tested. GF-3566 was used at 3.0, 3.5, and 4.0 pt/A and GF-2687 was used at 0.72 oz/A. Commonly used combination treatments such as (2, 4-D, MCPP and dicamba), SF formulation of (2,4-D, MCPP and dicamba) were included along with 2,4-D Ester alone. Several four-way combination treatments of (2,4-D, MCPP, dicamba and carfentrazone), (2,4-D, dicamba, quinclorac and sulfentrazone) and (2,4-D, dicamba, quinclorac, sulfentrazone and pyrimisulfan) were in some trials.

In general, most of the treatments provided excellent control of many of the broadleaf weeds present. The combination of halauxifen-methyl, fluroxypyr and 2,4-D choline (GF-3566) gave 90 to 98% control of common dandelion, white clover, broadleaf plantain, buckhorn plantain, henbit and yellow woodsorrel. Very similar control was obtained with the combination of halauxifen-methyl and florasulam (GF-2687). The treatment combination of halauxifen-methyl, fluroxypyr and dicamba at 4 pt/A resulted in similar control of many of the broadleaf weeds. Treatments of four-way combinations resulted in excellent control of most of the broadleaf weeds present in two of our trials. Our results indicate several alternative options for POST control of many broadleaf weeds in cool-season turfgrass.

LONG-TERM IMPACT OF NITROGEN, IRON SULFATE AND PLANT GROWTH REGULATORS ON ANNUAL BLUEGRASS POPULATIONS. K. TANG*, Penn State University, University Park, PA

ABSTRACT

Annual bluegrass (*Poa annua* L., ABG) is considered an undesirable species on golf courses in many regions of the United States. Numerous cultural and chemical practices have been studied for their effects on reducing ABG populations and promoting the desirable turfgrass species. However, the long-term effect of the combination of these practices is not sufficiently investigated. The objective of this study was to determine the long-term effect of nitrogen (N), iron sulfate (Fe) and plant growth regulators (PGRs) on ABG populations in a creeping bentgrass (*Agrostis stolonifera* L., CBG) golf green. The study was initiated on a mixed stand of 'L-93' CBG (75%) and ABG (25%) green at the Joseph Valentine Turfgrass Research Facility in University Park, PA from 2010 to 2018. Treatments were approximately every 3 weeks from May to Oct. Treatments included ammonium sulfate (24 or 147 kg N ha^{-1} yr^{-1}), iron sulfate (0, 12, or 49 kg Fe ha^{-1} $application^{-1}$), and the PGRs trinexapac-ethyl (TE, 0.043 kg a.i. ha^{-1} $application^{-1}$), flurprimidol (FL, 0.28 kg a.i. ha^{-1} $applicaation^{-1}$) or no PGR. Percent ABG was recorded every month during the growing season, using a 0.9 m × 1.8 m rating grid with 253 intersections. Annual bluegrass populations were influenced by N rate, Fe rate and PGRs. Lower N rate generally resulted in lower ABG populations. Iron applications initially resulted in lower ABG populations. Plots treated with flurprimidol generally had the lowest ABG populations and few differences existed between TE and applying no PGR. Finally, seasonal variation in ABG populations were minimal.

SMOOTH CRABGRASS CONTROL PROGRAMS WITH CHELATED IRON WEED CONTROL PRODUCTS. D.P. Tuck* and M.T. Elmore, Rutgers University, North Wales, PA

ABSTRACT

Field experiments were conducted to evaluate the efficacy of chelated iron containing herbicides for PRE and POST smooth crabgrass (*Digitaria ischaemum*) control at various rates, timings, and sequences in 2018 and 2019 in North Brunswick and Freehold, NJ, respectively. Both sites were a simulated lawn consisting of a poor stand of perennial ryegrass with history of smooth crabgrass. In 2018, treatments consisted of single or sequential applications of FeHEDTA (21 or 42 kg ha^{-1}), sodium ferric EDTA (35 or 56 kg ha^{-1}), corn gluten hydrolysate, and corn gluten meal (24 or 98 kg ha^{-1}). In 2019, treatments consisted of single applications or various sequential programs of FeHEDTA (10, 21, or 42 kg ha^{-1}), sodium ferric EDTA (35, 56, or 77 kg ha^{-1}), iron sulfate heptahydrate (10 kg ha^{-1}), corn gluten hydrolysate, and mesotrione (175 g ha^{-1}).

In both years, treatments were arranged in a randomized complete block design with four replications and applied to 0.9 by 2.0 m plots. Sodium ferric EDTA and corn gluten meal were granular formulations and applied using a shaker jar. All other treatments were sprayable and applied using a CO_2-powered single AI9504EVS nozzle boom with a water carrier of 820 L ha^{-1} and 1020 L ha^{-1} in 2018 and 2019, respectively. Initial applications were applied PRE on 12 June 2018 and 8 April 2019, respectively, and were irrigated immediately following application. Sequential POST applications of sodium ferric EDTA were applied to dew covered turf. Due to the delayed start of the 2018 experiment, emerged smooth crabgrass was controlled with fenoxaprop. Smooth crabgrass control was evaluated visually on a 0 (i.e., no control) to 100 (i.e., complete control) percent scale relative to a non-treated control.

2018 results demonstrated that FeHEDTA and sodium ferric EDTA provided substantial crabgrass control when applied PRE and early-POST, but the contribution of PRE or POST activity efficacy was not clear. In 2019, treatments that included sequential POST applications of FeHEDTA or sodium ferric EDTA generally provided greater smooth crabgrass control than treatments applied PRE. Sodium ferric EDTA applied PRE on 8 April, and POST on 23 May, and 17 June provided greater control (>75%) than sodium ferric EDTA applied PRE on 8 April and POST on 9 May (<30% control), at 13 WAIT. FeHEDTA applied POST on 23 May and 17 June provided greater control (48 to 60%) than FeHEDTA applied PRE on 8 April and POST on 9 May (<10% control), at 11 WAIT. Corn gluten heptahydrate and mesotrione applied singly PRE on 8 April, and iron sulfate heptahydrate applied PRE on 8 April and POST on 25 April, 23 May, and 17 June provided poor control (<30%) at 13 WAIT. FeHEDTA at 42 kg ha^{-1} applied singly PRE on 8 April or POST on 23 May, respectively, provided poor control for the duration of the experiment. These data suggest that PRE activity of FeHEDTA and sodium ferric EDTA is limited but sequential POST applications provide effective smooth crabgrass control. These chelated iron products generally provided better control when applied POST in June compared to May. Future research should investigate ideal application timings of chelated iron products for smooth crabgrass control.

SMOOTH CRABGRASS AND GOOSEGRASS CONTROL ON CREEPING BENTGRASS GREENS.
S.D. Askew* and J. Brewer, Virginia Tech, Blacksburg, VA

ABSTRACT

Creeping bentgrass maintained at greens height is typically more sensitive to herbicides. In the transition zone, crabgrass and goosegrass are problematic weeds, especially on greens that are experiencing biotic or abiotic stress. There currently aren't any postemergence herbicides registered to control crabgrass or goosegrass on creeping bentgrass greens. Experimental results have shown that fenoxaprop can be applied at low doses weekly or biweekly on greens to control young weed seedlings as they emerge. Our objective was to evaluate this frequent-treatment, low-dose approach with fenoxaprop, topramezone, quinclorac, and siduron for crabgrass and goosegrass control and creeping bentgrass response. Both quinclorac applied weekly and biweekly at 140 and 280 g ai/ha, respectively, and fenoxaprop applied weekly and biweekly at 17 and 35 g ai/ha, respectively, injured creeping bentgrass 30 to 60% at 52 days after initial treatment. Siduron at 5.6 or 13 kg ai/ha applied weekly or biweekly, respectively, did not injure creeping bentgrass and topramezone applied at 1.5 to 6.1 g ai/ha caused transient injury that never exceeded 15%. Fenoxaprop controlled both crabgrass and goosegrass while quinclorac controlled only crabgrass. Siduron completely controlled crabgrass but controlled goosegrass approximately 50% by season's end. Topramezone completely controlled goosegrass but controlled smooth crabgrass approximately 60% by season's end. In additional studies, siduron completely controlled smooth crabgrass when applied biweekly at four rates between 3.3 and 13 kg ai/ha. By reducing the siduron rate of 13 kg ai/ha in the previous study to 3.3 kg ai/ha, an average golf course could lower potential economic burden from $13,335/yr to $3,150/yr. Siduron is currently the only herbicide included in these studies that can be legally used on golf greens. These data suggest siduron is a safe option for season-long control of crabgrass and suppression of goosegrass.

VARIABLE RESPONSE OF CREEPING BENTGRASS TO TOPRAMEZONE ACROSS SEASONAL APPLICATION TIMINGS IN ALABAMA AND VIRGINIA. C. Goncalves*, J. Brewer, S. McElroy, and S.D. Askew, Virginia Tech, Blacksburg, VA

ABSTRACT

Creeping bentgrass (*Agrostis stolonifera* L.) response to topramezone is highly variable and influenced by herbicide rates, tank mixtures, and environment. The objectives of this research were to evaluate topramezone (Pylex) injury at different application rates and timing in Alabama and Virginia. Field treatments included applications of topramezone alone at (5.2, 10.4, 15.6, or 20.8 g ai ha^{-1}) applied in April, May, June, or July in Alabama and May, June, July, or August in Virginia. Results from those studies indicate early-season application of topramezone was safe to creeping bentgrass at all rates used. Topramezone application at high rates (15.6 or 20.8 g ai ha^{-1}) in late-season caused greatest creeping bentgrass injury, likely due to the high temperatures that stressed the turfgrass during this period. In general, injury from topramezone applications increased from April to July in Alabama and from April to August in Virginia. We conclude that early-season applications of topramezone at (5.2, 10.4, 15.6, or 20.8 g ai ha^{-1}) and low rate (5.2 g ai ha^{-1}) in late-season could safely be used on creeping bentgrass.

ANNUAL BLUEGRASS (*POA ANNUA*) CONTROL IN COOL-SEASON TURF WITH MESOTRIONE AND AMICARBAZONE TANK-MIXTURES.

M.T. Elmore*, D.P. Tuck, A.J. Patton, and A.W. Thoms, Rutgers University, New Brunswick, NJ

ABSTRACT

Field experiments were conducted in 2018 and 2019 to evaluate tank-mixtures of amicarbazone and mesotrione for POST annual bluegrass (*Poa annua* L.) control in Kentucky bluegrass (*Poa pratensis* L.) or perennial ryegrass (*Lolium perenne* L.) turf. In 2018 research was initiated at Rutgers University (Freehold, NJ) on 26 April. Treatments consisted of three amicarbazone-only programs and four amicarbazone + mesotrione tank-mixture programs. Amicarbazone-only programs consisted of three sequential applications of amicarbazone at 53 or 70 g ha^{-1} on 14-d interval or two applications on a 21-d interval at 110 g ha^{-1}. Amicarbazone + mesotrione programs consisted of three sequential applications of amicarbazone (54 g ha^{-1}) + mesotrione at 110, 140, and 175 g ha^{-1} on a 14-d interval and two applications of amicarbazone (54 g ha^{-1}) + mesotrione at 280 g ha^{-1} on a 21-d interval). Three sequential applications of methiozolin (950 g ha^{-1}) were included for comparison. In 2019, replicate experiments were conducted at Rutgers University, Purdue University (West Lafayette, IN) and Iowa State University (Ames, IA). Treatments consisted of amicarbazone (53 g ha^{-1}) applied alone or tank-mixed with mesotrione at 110, 140, and 175 g ha^{-1}. Tank-mixtures of amicarbazone (53 ha^{-1}) + mesotrione (110 g ha^{-1}) and UAN (30-0-0, 6.6 L ha^{-1}) or urea (1.2 kg N ha^{-1}) + AMS (2.6 kg N ha^{-1}) were also evaluated. Mesotrione alone at 175 g ha^{-1} was included for comparison. All treatments were applied three times on a 14-d interval beginning in late April or early May.

All sites were Kentucky bluegrass or perennial ryegrass naturally infested with annual bluegrass maintained at a 1.3 to 3.5 cm height. Treatments were arranged in a randomized complete block design, replicated four times, and applied using standard CO_2-powered small-plot spray equipment with water carrier (420 to 840 L ha^{-1}) and NIS (0.25% v/v). Annual bluegrass cover was evaluated visually on a 0 to 100 percent scale on the day of initial herbicide application and periodically thereafter until early autumn. Annual bluegrass cover was transformed and expressed a percent reduction (or increase) in cover compared the initial application date on a plot-by-plot basis. A grid intersect count was used to determine annual bluegrass cover at the final evaluation in early autumn. Kentucky bluegrass injury was evaluated visually when it was apparent. The ANOVA was conducted in SAS and Fisher's Protected LSD was used to separate means.

In 2018 all amicarbazone + mesotrione tank-mixtures reduced annual bluegrass cover more than all other treatments on the final rating date. Three applications of amicarbazone + mesotrione resulted in an 83 to 94% annual bluegrass cover reduction, compared to a 22 to 102% increase in annual bluegrass cover for amicarbazone alone and methiozolin. In 2019, responses were not consistent across locations. In New Jersey mesotrione + amicarbazone reduced annual bluegrass cover more than amicarbazone alone on most evaluation dates and at the conclusion of the experiment. In Indiana, mesotrione + amicarbazone reduced annual bluegrass cover more than amicarbazone or mesotrione alone at 6 WAT. But at the conclusion of the Indiana experiment all treatments reduced annual bluegrass cover similarly (by 70 to 80%). In Iowa, all treatments reduced annual bluegrass cover similarly (by >80%) at 6 WAT and at the conclusion of the experiment.

IRRIGATION TIMING INFLUENCES GOOSEGRASS CONTROL AND BERMUDAGRASS RESPONSE TO TOPRAMEZONE PROGRAMS. J. Peppers*, J. Brewer, J. Craft, and S.D. Askew, Virginia Tech, Blacksburg, VA

ABSTRACT

Lack of effective postemergence herbicide options makes goosegrass control more difficult in warm season turfgrass. The HPPD-inhibiting herbicide topramezone and the PSII-inhibiting herbicide metribuzin can both control goosegrass. However, injury concerns limit the use of these herbicides in warm season turfgrass. Previous research has shown that tank mixtures of topramezone and metribuzin reduces bermudagrass bleaching and allows for lower topramezone use rates to achieve effective goosegrass control. Further studies have recently shown that irrigating turf immediately after applying these herbicides may result in less turfgrass injury. Field trials were conducted during the summers of 2018 and 2019 in Blacksburg, Virginia, to examine the effect of irrigation timing on herbicidal activity of topramezone applied alone and in combination with metribuzin for goosegrass control and 'Latitude 36' bermudagrass injury. Treatments included topramezone (12.2 g ai ha^{-1}) alone and topramezone (6.1 g ai ha^{-1}) mixed with metribuzin (280.2 g ai ha^{-1}). Both treatments were examined with no irrigation, immediate irrigation (0.25 cm), irrigation 15 minutes after herbicide application, and irrigation 30 minutes after herbicide application.

Treatments that included topramezone plus metribuzin were significantly less injurious to bermudagrass than treatments of topramezone alone. For example, in 2018 at 7 days after treatment, immediate irrigation following topramezone alone injured bermudagrass 76% whereas similar irrigation following the combination of topramezone and metribuzin injured bermudagrass 5%. Turf response to herbicides and irrigation was dependent on year because irrigation at 15 and 30 minutes after treatment reduced phytotoxicity more in 2019 than in 2018. Irrigation, regardless of timing, reduced bermudagrass injury from topramezone plus metribuzin. At 28 days after application, topramezone and topramezone plus metribuzin controlled goosegrass at least 93% when irrigation was not applied following treatment. Immediate irrigation reduced goosegrass control to 57% and irrigation at 15 or 30 minutes reduced goosegrass control to approximately 81%. At 8 weeks after treatment, goosegrass plant counts exhibited a general trend toward increasing plant recovery as irrigation delay was reduced, but variable goosegrass population levels made statistical separation impossible. In 2018, the two herbicide programs did not differ for goosegrass recovery but topramezone plus metribuzin resulted in slightly more goosegrass recovery in 2019. Results from these trials indicate that more immediate irrigation lowered both bermudagrass injury and initial goosegrass control. Long-term goosegrass control may be governed by other factors.

EFFECT OF SOIL MOISTURE ON HERBICIDE EFFICACY FOR POSTEMERGENCE GOOSEGRASS (*ELEUSINE INDICA*) CONTROL. B.C. McNally*, M.T. Elmore, J.T. Brosnan, A. Shekoofa, D.P. Tuck, and J. Vargas, Rutgers University, New Brunswick, NJ

ABSTRACT

The objective of this research was to determine the efficacy of commonly used herbicides for POST goosegrass (*Eleusine indica* L. Gaertn) control under three soil volumetric water content (VWC) levels. Replicate experiments were conducted in a glasshouse at Rutgers University (New Brunswick, NJ) and the University of Tennessee (Knoxville, TN). Treatments consisted of five herbicides across three soil moisture levels arranged in a complete factorial. Herbicides consisted of fenoxaprop-p-ethyl (hereafter fenoxaprop; 140 g ha^{-1}), topramezone (25 g ha^{-1}), foramsulfuron (44 g ha^{-1}), 2,4-D + dicamba + MCPP + carfentrazone (860 + 80 + 270 + 28 g ha^{-1}), and thiencarbazone-methyl + foramsulfuron + halosulfuron-methyl (22 + 45 + 69 g ha^{-1}). These herbicides were applied to goosegrass subjected to low <12% VMC, medium 12-20% VMC, or high >20% VMC and replicated six times in a randomized block design. Goosegrass was planted from seed and allowed to establish in pots filled with field soil and fritted clay in a 4:1 ratio under well-watered conditions. Soil moisture treatments were initiated once goosegrass plants reached the three to six-tiller stage and continued for 20 days until herbicides were applied. Soil VWC was measured daily using a soil moisture probe and corrective irrigation was applied via watering can if needed. Herbicide treatments were applied using a research grade track sprayer (Generation III, DeVries Manufacturing) calibrated to apply 412 L ha^{-1} through a single 9502EVS nozzle; adjuvants were applied based on label recommendations. Goosegrass control was evaluated visually on a 0 (no control) to 100 (complete control) percent scale relative to the non-treated at 22, 34, and 43 days after herbicide treatment (DAT). Dry shoot biomass was measured at the end of the experiment (43 DAT) and transformed to a percent of the non-treated control (NTC).

At the Rutgers location, interactions of herbicide treatment and soil moisture were detected for all evaluations. With the exception of 2,4-D + dicamba + MCPP + carfentrazone, all herbicides provided better goosegrass control and the greatest biomass reduction to plants that were maintained at high VWC compared to low VWC at 43 DAT. Fenoxaprop and thiencarbazone-methyl + foramsulfuron + halosulfuron-methyl reduced biomass more at high VWC treatment than medium VWC treatment. 2,4-D + dicamba + MCPP + carfentrazone provided poor control regardless of VWC. Shoot biomass of plants treated with fenoxaprop and maintained at high, medium, and low VWC was 26, 61, and 81% of the NTC, respectively. Similarly, shoot biomass of plants treated with thiencarbazone-methyl + foramsulfuron + halosulfuron-methyl and maintained at high, medium, and low VWC was 27, 87, and 95% of the NTC, respectively. Results were similar in Tennessee as control was significantly greater for all herbicides when applied to goosegrass plants maintained at higher VWC.

Additional research was conducted at the University of Tennessee to explore the impact of high vapor pressure deficit (VPD; > 3 kPa) and temperature (32 and 38 °C) on goosegrass grown in two different soil types (silt-loam and sand) treated with foramsulfuron. Under high VPD and 38 °C, only treated goosegrass plants growing in silt-loam showed lower transpiration beginning 14 DAT. This experiment helped validate that foramsulfuron is more effective when goosegrass is in an environment that the air is dry (> 3kPa) and temperature is high. Also, plants grown in soil with higher water holding capacity (i.e., silt-loam) had greater transpirable soil water, which may be important for foramsulfuron efficacy.

DROPLET SIZE AND VELOCITY INFLUENCES SPRAY PENETRATION INTO ZOYSIAGRASS TURF CANOPIES. S.D. Askew* and J. Craft, Virginia Tech, Blacksburg, VA

ABSTRACT

Zoysiagrass can be treated with nonselective herbicides during winter dormancy for economical weed control. Previous research has shown that zoysiagrass injury from products like glyphosate are strongly associated with the amount of green tissue in the upper turf canopy. Although zoysiagrass may not have green leaves in the upper canopy, sporadic green leaves and stem tissue can be found in the lower canopy almost anytime during winter dormancy. Turf managers are perpetually fearful of the potential for turf injury when applying nonselective herbicides over zoysiagrass. Spraying systems technology has advanced in the past 30 years allowing for a large range of spray droplet characteristics. As droplet diameter and velocity increases, the resultant kinetic energy should increase droplet penetration into turf canopies and vice versa. Our objectives were to design a high-speed video analysis method to measure droplet diameter and velocity and determine spray parameters that minimize droplet penetration into lower zoysiagrass canopies. Using a completely randomized design, Teejet 11006 flat fan or turbo TeeJet induction (TTI) nozzles were operated at 103 or 414 KPa and at 25 or 61 cm above 10-cm diameter 'Meyer' zoysiagrass plugs. The resulting 8 treatments of the 2 x 2 x 2 factorial design were replicated 4 times. A 50% spray pattern indicator solution (Bull's Eye) was applied at 374 L/ha. Clippings were collected from the upper, middle, and lower third of each 5-cm tall canopy and dye was extracted with water and subjected to spectrophotometric analysis. Droplet characteristics for each of the 8 treatments were determined by high-speed video at 15,000 frames per second and 100 droplets were tracked across frames to measure diameter and velocity. As pressure increased, average droplet diameter decreased from 384 to 265 μm for flat fan and 735 to 651 μm for TTI. Droplet velocity increased from 14 to 29 km/hr as TTI nozzles were operated at 103 and 414 KPa, respectively. When averaged over all nozzles and pressures, droplet velocity increased from 17 to 25 km/hr as distance to target was decreased from 61 to 25 cm. These changes in droplet characteristics did not, however, significantly influence measured spray penetration based on our dye extraction technique. In all cases, approximately 70% of spray droplets were captured by the upper canopy, 19% by the middle canopy, and 11% by the lower canopy. Canopy biomass was consistent between treatments and thus, not a contributing factor. Future work will focus on improving consistency of dye recovery by identifying and removing potential sources of error. At this point, it is too early to make conclusions but the data are suggesting that substantial spray may reach the lower canopies of zoysiagrass regardless of nozzle type, pressure, or boom height.

BIG CHANGES FOR THE IR-4 PROJECT ON THE HORIZON. D. Kunkel* and J.J. Baron, Rutgers University, Princeton, NJ

ABSTRACT

After 56 years at Rutgers University, IR-4 Project Headquarters will be relocating, over the next two years, to a new host institution, NC State's College of Agriculture and Life Sciences. This move will ensure the long-term viability of The IR-4 Project with strong strategic alignment to that of the NC State College of Agriculture and Life Sciences' regulatory sciences and agriculture research.

The USDA-NIFA, IR-4 Project continues to successfully address pest control needs for specialty crop growers. In 2019, EPA approved nearly 1,600 new uses based on IR-4 research. Over 450 of those new uses were for herbicide label expansions on products such as trifluralin, *S*-metolachlor, bentazon, pyraflufen-ethyl, indaziflam, pendimethalin and others. New registrations and new uses were obtained that provide enhanced pest control in a broad range of specialty crops including tropical fruits, herbs and spices, leafy vegetables, root vegetables, tree fruits and others. IR-4 expects a similar number of approvals for 2020, with several submissions pending at EPA including products such as saflufenacil, clopyralid, pronamide, isoxaben, and quizalofop.

IR-4 also supports a significant amount of crop safety work to demonstrate that the new uses will not result in liability concerns. In 2018, IR-4 also initiated a new program to better serve the needs of the IR-4 stakeholders. This program, entitled Integrated Solutions (IS) Research Program, is designed to address pest problems without solutions, resistance management, products for organic production and pesticide residue mitigation. IS projects relating to weed control include: addressing the growing problem of Branched broomrape (*Orobanche ramose*, aka *Phelipanche ramose*) control in tomato, screening herbicides for organic weed control and weed control in brassica vegetables.

OVERLAPPING *S*-METOLACHLOR TREATMENTS IN PUMPKIN. K.M. Vollmer*, K. Nichols, D. Lingenfelter, M.J. VanGessel, B.A. Scott, and J. Wallace, University of Maryland, Queenstown, MD

ABSTRACT

Pumpkin (*Cucurbita pepo* L.) production in the northeastern United States is estimated to be about 10,000 ha. The onset of herbicide-resistant weeds in the area has limited herbicide options for postemergence control. Overlapping herbicides is a technique that involves sequential applications of soil-applied residual herbicides in order to overlap the herbicide's activity before the first herbicide dissipates. Separate field and greenhouse studies were conducted to evaluate weed control and pumpkin response to *S*-metolachlor applied as an overlapping residual treatment.

The greenhouse study evaluated the response of six pumpkin varieties to *S*-metolachlor applied to emerged pumpkin plants. The study was repeated. Application rates for the first run consisted of a 1X, 2X, and 4X rate, while the second run consisted of a 2X, 4X, and 6X rate. The 1X rate of *S*-metolachlor was 801 g ha^{-1}. Pumpkin injury developed within 7 days after application as leaf burn on leaves fully expanded at time of application. Up to 20% injury occurred within 7 d of application. Although rate and varietal differences were observed, they were not consistent across both trials. Injury symptoms were not observed on leaves that developed after the application.

The field study evaluated weed control efficacy and pumpkin response to *S*-metolachlor applied POST at different rates and application timings. The study was repeated at 3 locations in Delaware, Maryland, and Pennsylvania. Ethalfluralin was applied at 1,262 g ha^{-1} over the entire study immediately after planting. *S*-metolachlor was applied as a broadcast treatment at 801 or 1,602 g ha^{-1} 2, 3, or 4 wk after planting. An untreated control and a weed-free check was used of comparison. No injury was observed following POST *S*-metolachlor applications. *S*-metolachlor applied at 1,602 g ha^{-1} controlled both pigweed species and large crabgrass, 90%, compared to *S*-metolachlor applied at 801 g ha^{-1} (pigweed-82%, large crabgrass-83%) regardless of POST application timing.

An average of 8,645 pumpkins ha^{-1} were harvested, but there was no difference in location, *S*-metolachlor rate, or application timing among herbicide treated and weed-free plots. Pumpkin weight averaged 5 kg, but there were no differences among treatments.

Our results show that *S*-metolachlor does not cause any adverse crop injury when applied as a broadcast treatment in pumpkin. While this tactic did provide late season weed control, growers should be aware that *S*-metolachlor will not control emerged weeds. Therefore, additional strategies will need to be included to manage escaped weeds.

SUMMER SQUASH AND CUCUMBER RESPONSE TO RATE AND TIMING OF APPLICATION WITH *S*-METOLACHLOR. T.E. Besancon, B.L. Carr*, M. Wasacz, C. Austin, and E. Hitchner, Rutgers University, Chatsworth, NJ

ABSTRACT

In New Jersey, 3,561 ha of cucurbits were planted in 2018, the largest acreage of all fresh market vegetable commodities. Cucumber and summer squash are the two major cucurbit crops of the Garden State with a combined production value of $13 million. Due to many cucurbit varieties being low weed competing vines and their relative lack of tolerance to many herbicides, weed control often remains challenging. The number of currently labeled herbicides registered for use in cucumber and summer squash remains limited, emphasizing the importance of screening and evaluation of herbicides already registered for use on other minor crops. Field studies were conducted in 2018 and 2019 in Bridgeton, NJ, to test summer squash and cucumber tolerance to *S*-metolachlor applied PRE at planting or delayed PRE after crop reached the 2-leaf stage. 'Goldprize' summer squash and 'Python' cucumber were seeded prior to herbicide application. Treatments consisted of two standard labeled herbicides (clomazone + ethalfluralin at 0.88 kg a.i. ha^{-1}, and bensulide at 4.5 kg a.i. ha^{-1}), *S*-metolachlor applied PRE at 0.71 or 1.42 kg a.i. ha^{-1} alone or combined with bensulide at 4.5 kg a.i. ha^{-1}, and bensulide PRE at 4.5 kg a.i. ha^{-1} followed by a *S*-metolachlor delayed PRE at 0.71 or 1.42 kg a.i. ha^{-1}. A non-treated weedy check and non-treated weed-free check were also included. Crop stand counts were evaluated 4 weeks after planting (WAP). Crop stunting was rated 2, 4, and 6 WAP. Weed control was assessed 3 weeks after treatment (WAT), and crop yield weight was assessed according to USDA grades and standards 5 times in cucumber and 6 times in summer squash. *S*-metolachlor provided excellent (>98%) control of smooth pigweed (*Amaranthus hybridus*), eastern black nightshade (*Solanum ptychanthum*), carpetweed (*Mollugo verticillata*), and hairy galinsoga (*Galinsoga quadriradiata*). In summer squash, stand counts and yield were not affected by *S*-metolachlor, regardless of rates or timing of application, and all stunting symptoms were unnoticed 6 WAP. In cucumber, *S*-metolachlor decreased crop emergence up to 42%. Persistent crop stunting by 6 WAP was noted in plots where *S*-metolachlor was applied PRE alone (9% averaged on rates), or delayed PRE with 7% and 17% for 0.71 kg a.i. ha^{-1} and 1.42 kg a.i. ha^{-1}, respectively. Significant reduction in cucumber yield where noted with *S*-metolachlor applied PRE at 1.42 kg a.i. ha^{-1} (44% on average) or delayed PRE at 0.71 kg a.i. ha^{-1} (-34%) or 1.42 kg a.i. ha^{-1} (-50%). Overall, our results support the case for obtaining a New Jersey 24(c) Special Local Need label for summer squash but not for cucumber.

IMPROVING WATERMELON PLANT ESTABLISHMENT WITH RHIZOBACTERIA. M.A.
Cutulle*, Clemson University, Charleston, SC

ABSTRACT

Research was conducted to evaluate the impact of the rhizobacteria product SC-27 on watermelon vigor and fusarium resistance in greenhouse and field trials. Six watermelon varieties (Charismatic, Fascination, Sugar Baby, Tri-X-313, Charleston Gray, and USVL252FR2) were treated with 4 different inoculation treatments: SC27 soil microbe inoculation, FON2 inoculation, SC27 + FON2, and no inoculation. Numerically, the more vigorous plants were treated with SC-27 while the least vigorous plants were not treated with SC-27. Generally, not many differences among variety on vigor in greenhouse trials; however, fascination tended to be the most vigorous variety when averaged across inoculation treatments. In the field SC-27 applications did not significantly affect the number of melons produced for each variety for total harvest. Interestingly, the effect of SC-27 was observed more on the 2nd harvest than the 1st harvest. The study needs to be repeated before one can conclude SC-27 increases yield.

FIELD PEA (*PISUM SATIVUM*) AS A SUMMER COVER CROP IN NEW ENGLAND VEGETABLE PRODUCTION SYSTEMS. R.N. Brown*, URI, Wakefield, RI

ABSTRACT

Climate change and demand for locally grown produce are leading more farmers in southern New England to grow fall crops of broccoli and cool season greens. This creates an opening for a summer cover crop to suppress weeds and provide nitrogen for the fall vegetable crop. Field pea is a traditional summer annual cover crop in northern New England and could be suited to late spring and early summer use in southern New England. The objective of this study was to evaluate a range of seeding dates for field pea as a cover crop in southern New England.

The trials reported here were conducted at the University of Rhode Island's Gardiner Crops Research Center in Kingston, RI. In 2016 'Austrian Winter' peas were seeded at 112 kg/ha on April 15, April 29, May 13 and May 26. In 2018 '4010' forage peas were seeded at 112 kg/ha on May 8, May 22, June 5, June 19, July 3 and July 17. In 2019 '4010' peas were seeded at 112 and 224 kg/ha on May 7, May 21, June 4, June 18, July 2 and July 16. Fields were prepared by disking and peas were planted with a grain drill set to 1.3 cm depth. Total planting size varied across years, but all data were collected using randomly located 2 ft x 2 ft (0.6 m x 0.6 m) subplots within the planted area. Pea biomass and weed biomass were measured at six weeks after seeding, and pea biomass was measured at eight weeks after seeding. All vegetation within the quadrat was clipped within 3 cm of the soil surface. Peas were separated from weeds, and samples were dried and weighed.

In 2016 the May 26 seeding significantly out-yielded the other seeding dates at both 6 weeks and 8 weeks after planting (Figure 1). In 2017 the June 19 seeding yielded the most crop biomass at 6 weeks after seeding, but only May 8 and July 3 were statistically lower (Figure 2). The May 22 seeding yielded the most crop biomass at 8 weeks, with only July 3 and July 17 statistically lower (Figure 2). Crop biomass exceeded weed biomass on all dates except July 3 and 17. In 2019 the June 4 planting produced the most biomass at 6 weeks at both seeding rates (Figures 3 and 4). At the 100 lb/acre seeding rate the May 21 seeding date produced the most biomass at 8 weeks, significantly more than any other date (Figure 3). At 200 lb/acre the May 21 and June 4 dates were nearly identical, and both significantly out-yielded the other dates (Figure 4). The 200 lb/acre seeding rate produced significantly more biomass than the 100 lb/acre rate on all dates except for the 6-week sampling of the May 7 planting. The May 7, May 21 and June 7 plantings had significantly less weed biomass than the three later plantings. For the June 18, July 2 and July 16 seeding dates weed biomass exceeded crop biomass at the 100 lb/acre seeding rate, and was similar to crop biomass at the 200 lb/acre seeding rate.

Field peas appear to be a good option for spring to summer cover cropping before fall vegetables in southern New England, with May and June being the best months for planting. Allowing the peas to grow for 8 weeks significantly increases biomass over termination at 6 weeks. Increasing seeding rate over 100 lb per acre significantly increases biomass, but that may be a result of generally poor stand establishment. More research is needed to determine the cause of poor stands, and improve percent establishment.

CRANBERRY CROP RESPONSE TO DIFFERENT APPLICATION METHODS OF SULFENTRAZONE AND FLUMIOXAZIN. K. Ghantous, T.E. Besancon, and H.A. Sandler*, UMass Cranberry Station, East Wareham, MA

ABSTRACT

During recent screening trials, sulfentrazone and flumioxazin demonstrated crop safety when applied to dormant cranberry plants and control of high priority weeds such as dodder (*Cuscuta* spp.) and haircap moss (*Polytrichum commune*). Both herbicides are PPO inhibitors in WSSA Group 14/HRAC Group E, which would be a novel mode-of-action group for cranberry.

Flumioxazin studies in WI resulted in crop injury not seen in MA or NJ. Cranberry growers in WI typically apply pesticides with boom sprayers (187 L water ha^{-1}). MA growers apply through the irrigation system (chemigation; approximately 3,740 L water ha^{-1}), while NJ growers use both methods. Both sulfentrazone and flumioxazin are soil-active compounds that require rainfall or irrigation for activation. Herbicides applied by boom sprayer are more concentrated and may be in contact with crop foliage longer than chemigated herbicides, which are more dilute. The effect of herbicide application method on cranberry crop safety for these two herbicides has not been previously reported. Through our observations, we hypothesized that irrigation immediately after boom application (i.e., wash-off) could minimize the time that herbicides were in contact with crop foliage without decreasing efficacy.

Experiments were conducted in two cranberry growing regions, MA and NJ, to study the effects of application method on crop safety. Sulfentrazone (39.6% a.i.) at 840 g ha^{-1} or flumioxazin (51% a.i.) at 420 g ha^{-1} in MA or 560 g ha^{-1} in NJ was applied as a simulated boom application (187 L water ha^{-1}), boom application followed by (fb) wash-off (187 L water ha^{-1} followed by 0.25 cm water), or chemigation (3,740 L water ha^{-1}). For each herbicide, each plot received a single treatment at one of two timings based on cranberry phenology (spring dormant or cabbage head) and all treatments were replicated four times at each location.

For sulfentrazone applications in both MA and NJ, no treatment resulted in significantly less fruit than the untreated control, and the effect of application method was not significant. There was a trend towards lower yields in plots treated at the later timing than the earlier timing (not statistically significant).

For plots treated with flumioxazin at the 420 g ha^{-1} rate (MA), application method had no effect on yield. For the higher 560 g ha^{-1} rate (NJ), the effect of application method on yield varied by timing; all treatments reduced yield compared to the untreated plots except for the spring dormant timing fb wash-off. For MA and NJ, plots treated at the later timing had less fruit than plots treated at the earlier timing ($p = 0.033$ for MA and $p = 0.077$ for NJ), which did not differ from the untreated control.

Preliminary results suggest that 840 g ha^{-1} sulfentrazone can be applied to dormant cranberries without reducing yield. Flumioxazin caused injury at the higher 560 g ha^{-1} rate used in NJ, and at this rate, application method became significant. Our results indicate that application method may play an important role when herbicides are used at rates high enough to cause cranberry injury.

IMPACT ON CRANBERRY YIELD AND FRUIT QUALITY OF WEED DENSITY FOR FIVE COMPETING SPECIES. T.E. Besancon*, B.L. Carr, K. Ghantous, H. Sandler, and J. Colquhoun, Rutgers University, Chatsworth, NJ

ABSTRACT

An enhanced understanding of weed species and density impacts on cranberry yield and quality would allow for management prioritization, but surprisingly, minimal documentation of these relationships is available. This information would be particularly useful as growers consider reducing inputs to address economic challenges. They would be able to focus on those weeds that have the greatest crop impact as opposed to those that may be abundant but primarily affect aesthetics. With this in mind, our objectives were to: 1) establish a methodology to document weed impacts on cranberry yield and quality; and 2) test the methodology across a spectrum of weedy species and regions. The long-term goal is to establish a "library" of weed species thresholds that could serve as a resource for growers, crop scouts and other pest management decision makers. Quantified weed parameters included density, biomass and ground cover. Cranberry parameters included berry yield, number, color, rot, insect damage, total anthocyanin content (Tacy) and Brix. Weed species included dewberry (*Rubus* spp.) And slender-leaved goldenrod (*Solidago tenuifolia*) in Wisconsin, Carolina redroot (*Lachnanthes caroliniana*) and maple (*Acer* spp.) in New Jersey, and moss (*Sphagnum* spp.) In Massachusetts. dewberry negatively affected not only cranberry yield but also color. Each percent of dewberry ground cover reduced cranberry yield by two barrels (100-lb weight) per acre. Increases in goldenrod biomass and density were equally predictive of a negative response in cranberry yield. The relationship between Carolina redroot biomass and density were also negatively related to cranberry yield. Interestingly, a strong positive relationship was documented between Carolina redroot density and fruit insect damage. As might be expected, the impact of moss on cranberry yield and quality was more challenging to quantify as moss biomass harvest was difficult and other factors likely affect cranberry production in areas where moss was prevalent, such as moisture management. With that said, cranberry yield was still inversely related to moss biomass. This work will be repeated in 2020 and expanded to include additional weed species.

SYMPOSIUM: INCLUSION IN SCIENCE. Caren A. Schmidt and Elizabeth Mosqueda, BASF, Research Triangle Park, NC and California State University-Monterey Bay, Seaside, CA

ABSTRACT

The "Inclusion in Science" symposium was held on Tuesday, Jan 7, from 1:30-3:30pm. It included a keynote address by Dr. Elizabeth Mosqueda, Assistant Professor of Crop Management at California State University-Monterey Bay, documenting the impact of Diversity and Inclusion efforts that she led as a graduate student with the Western Society of Weed Science. Following the keynote, Dr. Mosqueda and Dr. Caren Schmidt (Vice-President, NEWSS) facilitated an interactive workshop exploring instances where our differences have impacted quality of life, understanding the definitions of diversity, benefits of inclusion, and what an inclusive environment should look like, exploring the ways in which implicit bias has impacted our work life, and creating action plans to become more inclusive in our daily lives. Following the Symposium, several members expressed interest in serving on the NEWSS Inclusion ad-hoc committee to explore/identify opportunities to be proactive in recruitment of underrepresented groups to our annual meeting and ways to create a more inclusive society and meeting environment to leverage our existing and future diversity.

NEWSS YEAR-END FINANCIAL STATEMENT 2019

AUDIT REPORT – Kurt Vollmer

Financial documents and bank accounts were audited by,
Quintin Johnson and Barbara Scott on December 17, 2019

BEGINNING STATEMENT: As of October 31, 2018, the NEWSS checking account showed a balance of **$61,783.14**, while the savings account balance was **$31,857.27**. The endowment fund balance was **$29,786.33**.

ENDING STATEMENT: As of October 31, 2018, the NEWSS checking account showed a balance of **$67,362.51**, while the savings account balance was **$31,863.62**. The endowment fund balance was **$34,902.05**.

Net gain in assets for the fiscal year: $134,128.18- $123,426.74 = $10,701.44

FISCAL YEAR ACCOUNTING

New income as shown on statements for November 2018 through October 2019 including earned interest and endowment gain was **$51,489.96**.

New deductions from checking for November 2018 through October 2019 totaled $**40,788.52**.

Net ***gain*** *in assets:* ***$51,489.96 – $40,788.52 = $10,701.44***

Non Annual Meeting Revenue/Expenses:

Non annual meeting revenue, which was comprised of interest earned as well as member dues only, sustaining members ($8,500), and other contributors totaled **$14,070.96**.

Non annual meeting expenses (Dishonesty bond, website, administrative costs, Dir. Science Policy (2019), National Weed Contest, WSSA abstract upgrade, 2020 Hotel, and endowment fund investment totaled **$20,807.88**.

Annual Meeting Revenue/Expenses:

Annual Meeting revenue (registration, industry social, proceedings residuals) totaled **$35,329.00**.

Annual meeting expenses (hotel + AV, PayPal fees, profit distribution, programs, awards, meeting management, honorariums, and 2020 hotel deposit) totaled **$19,980.64**.

Education Fund Contributions

Contributions received for the Education Fund include proceeds totaled **$2,090** for FY 2018/2019.

NEWSS FISCAL STATEMENT
NOVEMBER 1, 2018- OCTOBER 31, 2019

Income Received	Amount
Joint Meeting Total Registration	$26,850
Member Dues Only	$350.00
BASF awards	$300.00
Industry Social (4 Sponsors)	$8,005.00
Interest earned on savings account	$6.35
Net gain in endowment	$5,115.72
Sustaining Membership: 10 to date	$5,000.00
Sustaining Designated for Weed Contest	$3,500.00
Proceedings Residuals	$174.00
Endowment Contributions	$2,090.00
Refund: Hilton Penn's Landing	$98.89
Total NEWSS Income Nov 1, 2018- October 31, 2019	**$51,489.96**

Expenses Paid	Amount
PayPal Fees	$881.61
Joint Meeting Hotel + AV Cost	$11,543.58
Joint Meeting Profit Distribution 2019	$317.93
Sustaining Member Poster	$227.88
Alphagraphics - Mtg Programs	$522.86
EC Dishonesty Bond Insurance Renewal	$359.00
NEWSS/NEPPSC website	3,000.00
Website "A Small Orange" domain fee for NEWSS.org.	$228.00
Pastoor Honorarium	$2,248.01
Administrative	$147.07
WSSA Rep Travel to Summer Board Meeting	$952.16
Meeting Management	$2,429.32
Quiz Bowl-Amazon Gift Cards	$320.00
Annual meeting Membership Awards - Etch Art	$468.25
NEWSS Award Booklets	$271.20
Annual Meeting Poster Awards (1st and 2nd)	$300.00
Annual Meeting Student Paper Awards (1st and 2nd place)	$300.00
Annual Meeting Photo Contest (1st, 2nd and 3rd place)	$150.00
2018 Endowment Investment	$1,760.00
NEWSS Weed Contest Team Travel	$400.00
NEWSS Weed Contest Plaques	$1,724.56

NEWSS Contribution to NA Weed Science Contest	$5,000.00
WSSA Dir. Science Policy YEAR 2019 Dues	$2,400.00
WSSA Abstract Upgrade	$2,000.00
2020 Hotel Deposit	$2,500
2019 Fall Board Meeting	$337.09
Total NEWSS Expenses paid Nov 1, 2018- October 31, 2019	**$40,788.52**

Total Fiscal Year Summary

Total Income	$51,489.96
Total Expenses	$40,788.52
Net Gain	**$10,701.44**

NEWSS Account Balances October 31, 2019

Checking	$67,362.51
Savings	$31,863.62
Education Fund	$34,902.05
Total Net Assets	**$134,128.18**

NORTHEASTERN
WEED SCIENCE SOCIETY

2019 NEWSS BUSINESS MEETING
JANUARY 10, 2019

Call to Order

President Carroll Moseley called the meeting to order at 1706.

Minutes

Minutes of the 2018 Annual Business meeting were distributed for review.

Typos – Fix "Van Gessel", change "Jay" Craft to "Jordan" Craft.

Motion by Dwight Lingenfelter, second by Wes Everman. Minutes as amended approved unanimously.

Necrology Report – Thierry Besançon

George H. Bayer. Former NEWSS President, long-time scientist with Agway.

Executive Committee Reports

President, Carroll Moseley

Carroll described the duties and services provided by Bill Torres. He complimented the membership on their active participation in this year's meeting. Kudos to Dan Kunkel, Program Chair, and Theresa Piskackova, Graduate Student Representative for her efforts.

Treasurer's Report – Kurt Vollmer

Kurt briefly reviewed the Weed Contest, and praised BASF for the execution of the contest, and for covering all costs.

Kurt summarized current accounts. For 2018 we had a negative balance of -$574. He explained that full-fee registrations are down.

The 2018 Meeting brought in $32,205, and Kurt summarized the contributions to the meeting by Sustaining Members.

NEWSS Account Balances, 10/31/18	
Checking	$61,783.14
Savings	$31,857.27
Student Endowment Fund	$26,194.01
Total Net Assets	$123,046.74

Fiscal Year Summary	
Total Income	$53,829.65
Total Expenses	$54,404.19
Annual Balance	-$574.54

Archive Committee – Dan Kunkel

Shawn Askew provided his Presidential Library, via USB drive, to Dan. Dan reminded Randy Prostak that his Presidential materials are needed.

Membership Report – Thierry Besançon

Thierry summarized membership trends over the last 10 years. We are up this year from a low in 2018.

Awards Committee – Randy Prostak

Randy reviewed the Awards presented this year

NEWSS Fellow – Toni DiTommaso

Award of Merit – Matt Mahoney

Outstanding Researcher – Michael Flessner

Outstanding Educator – Hilary Sandler

Robert D. Sweet Outstanding Student, Ph.D. - Shawn Beam

Robert D. Sweet Outstanding Student, M.S. – Kara Pittman

M. Garry Schnappinger Service Recognition Award – Erin Hitchner

Reviewed Weed Contest Results (extract from Summer newsletter).

STUDENT PAPER AWARDS –Shawn Askew

Comments to contestants: tips for effective laser pointer use; need Hypothesis and Objectives; conclusions should be conclusive, not results; too many slides with small fonts; talks need to hit the 12-minute mark – several were short; slide colors were often not contrasting enough; and voice inflection as well as turning away from the microphone when looking at the screen.

Honorable Mention –Interaction of Organic Herbicide and Cover Crops, Danielle Lewis, Clemson University.

Second Place – Annual Bluegrass Weevil (*Listronotus maculicollis*) and Paclobutrazol for Annual Bluegrass (*Poa annua*) Control in Fairways, Katy Diehl, Rutgers University.

First Place – Determining Sorgoleone Efficacy on Select Wheat (*Triticum aestivum* ssp.) Cultivars and Weed Species, Eric Jones, North Carolina State University.

PHOTO CONTEST – presented by Shawn Askew

Second Place – Dwight Lingenfelter, musk thistle

First Place – Thierry Besançon, crimson clover

POSTER AWARD – presented by Matt Elmore

Honorable Mention – Evaluation of Weed Size and Water Carrier pH on Suppress Herbicide Efficacy, Danielle Lewis, Clemson University.

Honorable Mention – Critical Period of Weed Control for Ryegrass in Winter Wheat, Diego Contreras, North Carolina State University.

Second Place – Modeling Palmer Amaranth (*Amaranthus palmeri*) Emergence using Chronological, Thermal, and Hydro-thermal Time, Theresa Piskackova, North Carolina State University.

First Place – Bulbous Bluegrass: A Weed of Ornamental Turf, John Brewer, Virginia Tech.

Jeff Derr motioned, Toni DiTommaso seconded, reports accepted unanimously.

Officer Changeover

Carroll Moseley passed the Kilmer Oak gavel to now-President Dan Kunkel

New Business – Dan Kunkel

Dan Kunkel presented the traditional gavel plaque to Carroll Moseley for his service as President.

RESOLUTIONS COMMITTEE REPORT – Jerry Baron

Jerry described a resolution forwarded by Carroll Moseley, related to communications regarding glyphosate. At this time, the resolution will be further developed and offered for the 2020 meeting.

COMMITTEE APPOINTMENTS

Resolutions Committee – Mark VanGessel, Renee Keese, John Wallace.

Nominations Committee – Carroll Moseley, Toni DiTommaso appointed. Nominated – Sandeep Rana, Matt Elmore, and Shawn Askew. Neal moved, Keese second, passed unanimously.

EXECUTIVE BOARD APPOINTMENTS

Matt Cuttule was nominated to take over for Quintin Johnson as Research and Education Committee Chair, and Michael Flessner offered to stay on for one more year as Editor. Wes Everman motioned, Sudeep Matthew seconded, unanimously approved.

VICE-PRESIDENT ELECTION

Caren Schmidt was presented as the nominee for Vice President. Dan asked for nominations from the floor. Dwight Lingenfelter moved, Carroll Moseley seconded, to close nominations. Motion passed unanimously. Vote for Caren Schmidt was unanimous.

FUTURE MEETINGS

Dan shared that the Board is considering an invitation from WSSA to meet jointly in 2023. Board members will meet with the WSSA Board at the WSSA meeting to discuss.

Art Gover shared an overview of the 2020 program in Philadelphia.

ENDOWMENT COMMITTEE BYLAWS

Art Gover explained the ballot the Society received via email, asking for the Society membership to approve amending the Society Bylaws with the bylaws specific to the Endowment Fund. The Executive Committee approved the Endowment bylaws at the Board Meeting on January 7, 2019.

SCIENCE POLICY REPORT

Federal shutdown is not over.

WSSA is filling a legislative intern position.

Invasive Weed Awareness Week is coming up in February. WSSA will be developing a 'Weed Bingo' game, kidding not, to be released for the IWAW in 2020. This is essential, as the entomologists have a "Bug Bingo" game, and ours will be much better.

ENDOWMENT RAFFLE

Todd Davis is your winner of the 50/50 raffle to benefit the Endowment Fund. Todd donated $270 of the $420 pot to the Endowment Fund.

INTRODUCTION OF THE 2019 EXECUTIVE BOARD – Dan Kunkel

Carroll Moseley:	Past President
Dan Kunkel:	President
Art Gover:	President-elect
Kurt Vollmer:	Treasurer
Thierry Besançon:	Membership Chair
Kate Venner:	Public Relations Chair
Lane Heimer:	Editor
Matt Cuttule:	Research and Education Chair
Wes Everman	Sustaining Membership Chair
Theresa Piskackova:	Graduate Student Representative
Uriel Menalled:	Graduate Student Representative-elect
Rakesh Chandran:	WSSA Representative
Hilary Sandler:	CAST Representative

Adjourn

Kate Venner motioned, Thierry Besançon seconded, vote unanimous. Meeting adjourned at 1820.

Author Index